WHY
WOMEN
PIVOT

Congratulations!
Keep mentoring!
Pivoting!

[signature]

WHY WOMEN PIVOT

EMBRACING TRANSITIONS
UNIQUE TO OUR CAREERS

TRISH BARBER

MANUSCRIPTS
PRESS

COPYRIGHT © 2024 TRISH BARBER
All rights reserved.

WHY WOMEN PIVOT
Embracing Transitions Unique to Our Careers

| ISBN | 979-8-88504-346-5 | *Paperback* |
| | 979-8-88926-016-5 | *Ebook* |

In loving memory of my mother, who would shake her head at my audacity for writing a book.

Also dedicated to my husband, Steve, who keeps me from getting lost.

Life is not easy for any of us. But what of that? We must have perseverance and, above all, confidence in ourselves. We must believe that we are gifted for something and that this thing must be attained.

—MARIE CURIE, PHYSICIST, CHEMIST,
NOBEL PRIZE WINNER

Contents

AUTHOR'S NOTE 11

1. CAREER GROWTH DURING LIFE PHASES 19
2. WHY DO WE PIVOT? 37
3. NAVIGATING ORGANIZATIONAL CHAOS 57
4. IDENTITY 71
5. ENTREPRENEURIALISM 83
6. THE BUSINESS OF MOTHERING 103
7. UNEXPECTED WAYS OF IMPACTING PERSONAL BRAND 117
8. NETWORKING 133
9. WHAT COMPANIES NEED TO KNOW 149

CONCLUSION 163
ACKNOWLEDGMENTS 169
APPENDIX 173

Author's Note

I would tell my younger self to set high goals and not to be disillusioned if there isn't a hockey stick trajectory to your career. That's not the way this usually works for women. But the opportunities are endless if you can see where you want to go, create options, and take the long view.

We women are coming into our own. We are outpacing men in college enrollment and graduation rates and are a majority percentage of the US labor force. Research cited in this book shows that primary bread-winning mothers are increasingly becoming the norm. This should mean opportunities abound, but women face unique challenges due to the other unique roles we also hold.

Our biological roles of giving birth to children aren't changing. Pregnancy and childbirth require stamina and then recovery. We remain largely responsible for the *unpaid* work of caring for the physical and emotional needs of children and taking on the household chores that keep families functioning. We perform these roles despite hormones and exhaustion and while our bodies are physically changing. We also care for parents and other older relatives.

Not surprisingly, we are susceptible to confidence gaps because of voices in our heads and because we must move between these situations daily and seamlessly. These different roles require opposite skill sets and emotional states. It's a moving target because as children grow, we must learn how to parent more strategically and practice letting go while still growing in our careers. Our own parents age, and we support them through later-in-life transitions. Family composition may also change, and we can find ourselves navigating personal transitions while still being responsible for the well-being of others.

Given the paid and unpaid work that women do, it's worth taking a moment to define what I include in a woman's career. Most women will move in and out of the paid workforce during their lifetime. Since women are more likely to stay at home, either full time or part time to care for family, this choice to take on the unpaid work *is* a career choice. While it may be temporary, it could last for years, making reentry into the workforce a challenging pivot that comes with penalties. Our careers include the months and years we are home with family, and this book explores the pivots in and out of the job market. This is important when attitudes about working outside the home are changing. Now, research shows women prefer to work outside the home versus staying at home with family. COVID-19 is the latest large-scale example of when we pivoted home because the number of women in the workforce dramatically decreased.

Purging some demons was necessary for me here. The first draft of this manuscript was more raw. My entrepreneurial ventures and time at AOL coincided with me having young children and yielded many important lessons, both through example and counterexample. Voices from my upbringing in rural New York State as the oldest child in a patriarchal, working-class family also provide context for my personality and what I still choose to work on. For example, my background contributes to my deference to authority figures and my need to please others, which probably didn't serve me well in the early stages of my career. I believed I was lucky to be in those professional situations versus deserving. I also thought men were righter than women. I've since learned better.

I've also learned that being too quick to concede gets you nowhere—either personally or professionally. Being disruptive and holding firm takes practice, and the best role models I've had in this area were men. I cite compelling research about others' perceptions of women in roles of authority, saying we must straddle a line when asserting ourselves to achieve maximum impact. I also discuss how creating a culture of inclusion in the workplace does a lot to dispel any of these biases we bring into work situations.

When thinking about role models, I've added a few in this book. Senior-level women I've worked for haven't always been huge advocates for others, but they have taken on big roles and made huge contributions to the companies they represented. I've highlighted a few women who made mentoring and advocacy a priority. As Madeleine Albright said, "There's a special place in hell for women who don't help each other." With the best of intentions, and because there are fewer of us at the table, power dynamics can sometimes interfere with our natural instincts to take care of each other. Being aware of this will help us become better mentors and better leaders to all those on our teams.

It's not only my voice in this book. I've worked to capture advice from other women and third-party resources that is quickly actionable and provides ideas that may need to percolate. The span of our careers requires both kinds of guidance. Because pivots don't happen overnight, we can ponder thoughtful questions, do research, and talk to others while readying our plan.

Circumstances can sometimes make acting difficult or impossible. For example, in the throes of raising small children, women have less time for professional development activities but can certainly identify things to explore when time allows. Similarly, incorporating realism sets you up for success. An example of this for me is when I decided to prioritize exercise for stress relief and post-baby weight loss. I had no time or energy for regular gym visits. What was manageable was an occasional drop-in Jazzercise class. Once our kids got older and I was getting more sleep, regular gym visits became possible. Being overwhelmed to the point of doing nothing isn't a great place to sit, but finding small ways to do something helps us feel accomplished and creates an upward spiral.

Because we need to be both practical and strategic, the "Pivot Lessons You Can Use" section in each chapter summarizes important ideas that you can come back to again and again.

It compiles the practical tips and light bulb moments I wish I had when navigating my career, settling into a partnership, and eventually raising three children. It captures invaluable perspectives from wise women who seized opportunity or chose to defer opportunity for an important reason.

At certain points in my career, juggling family was extremely overwhelming, and I wouldn't have had the stamina to keep going if not for three amazing day care providers who supported our family at various phases. They were hard to find, and I've shared what I know about finding good providers here. I had one woman tell me she would have rejoined the workforce sooner after having her children, but a bad day care experience put her off for several more years. I will talk about the importance of developing a support network that includes friends, extended family members, various community members, sometimes colleagues, and, of course, day care providers. I want women to know there isn't one right way to do this, but you will need help. Whichever way you choose, you will reevaluate often. That's normal and necessary.

When writing this book, an experienced colleague said, "Write what you know." I could have written a book on the history and evolution of online marketing or a tell-all book about the early days of online media. However, that wasn't nearly as interesting to me as capturing perspectives of women balancing interesting opportunities with the unpredictable realities of life. After the many pivots in my career, I have a lot to share. The trauma of COVID-19 provided a personal tipping point.

COVID-19 changed things in ways we are just starting to understand. We had one child still living at home for her last two years of high school and attending online classes. There was extra stress associated with her taking SATs in this environment, and filling out college applications without visiting universities was challenging. Our daughter and all the other kids her age missed the normal high school rites of passage.

My siblings and I also lost our dad in May of 2020 while he was locked down in his nursing home. He never actually got COVID-19, but he succumbed to complications from lung cancer. They granted us an end-of-life visit to say goodbye, and it was horrifying to see the changes that took place over the ten weeks we weren't allowed inside. Dad was confused, nonverbal, had experienced significant weight loss, was poorly shaven, and was profoundly struggling to take a breath. Situations like this happened to everyone, and these traumas, combined with the hostile political climate, took a toll on us all. I wondered how I could do something positive in this environment.

During the pandemic, technology allowed colleagues to gather for meetings on video, and we got a new glimpse into each other's lives that changed the tone of the professional landscape. Everyone was a little more patient, supportive, and casual. While the mentoring and networking programs I was a part of before COVID-19 were shut down, I started writing to capture the things that women had wanted to know more about before COVID-19. Topics included female innovation, identity, self-perception, self-sabotage, personal branding, trade-offs, and options for becoming entrepreneurs.

What began as a passion project of gathering ideas that support women as they work through their careers helped me better understand my own pivots and gave me a reason to connect with other women to simulate missing networking and mentoring opportunities. In mid-2022 I realized I had the makings of a book, and I got serious about writing. Talking with women about how they found their power during a reflective time like a pandemic was inspiring. We always need smart women in our heads as we take on each phase of our careers. That's when writing the book outweighed the reasons not to.

I knew I would need to address each life phase in this book to properly reflect how goals and priorities shift. The first chapter lays out four unique phases: learning, accomplishing, nurturing, and reinventing. The rest of the book weaves ideas from these phases into the topic of each chapter and shows where pivots become a necessary part of growth—even the ones involving slowing your professional roll.

I've been married to my husband Steve for over thirty years, so my perspective is one of a traditional family with two full-time, working adults. We have had several pivots—some were out of necessity, and others were about opportunity or experimentation. You'll hear about Steve throughout the book. In our case, acquisitions, mergers, relocations, and reorganizations drove those pivots. They were also driven by raising a family of three children—one with a scary health issue in her tweens. Now that our children are becoming independent adults, that phase of reinvention is here.

Let's face it: This empty-nester or Reinvention phase is where we all get a little crazy. Everyone prioritizes what's left on our bucket lists. This book shares some of the questions that participants are asking themselves during this time since technology has opened so many new doors. People our age are healthier longer, and the gig economy makes it possible to contribute meaningfully without full-time commitment. That makes it possible for us to still experiment and explore. The options can be overwhelming. This book sorts through the triggered anxieties and the solutions others have implemented.

One final note, I'll also say that our self-esteem is wrapped up in our ability to contribute, both to our families and to society. A friend confided that her family treated her much more respectfully after she pivoted from being a stay-at-home mom to a career where she made a lot of money. Her family dynamic completely changed, as did her self-perception. This is not the only time I've heard a story like this. Economic independence correlates to respect from others and self-esteem.

My goal here was to empower women of all ages as they continuously reevaluate where they are and what they want next. This project has also allowed me to bring focus to two things I've long been involved with and passionate about: the importance of women in leadership and getting girls involved in STEAM.

I hope you will read on to learn more and confidently structure the career you have been dreaming about.

1.

Career Growth during Life Phases

———

To use a colloquialism, I'm most comfortable when there is some "spaghetti on the wall." That approach has kept life interesting and allows for experimentation. If things are too neat and orderly, I get bored. There's no fun when everything is already organized. Throughout my career, I've been good at recognizing and seizing opportunities, but I was reactive early on. Learning to map out a course of action came later. We just don't know enough at the beginning of our careers to do much planning beyond getting training, getting hired, learning an industry, applying what we've learned, and seizing opportunities. We are also figuring out who to trust, how to behave, where we fit in, and how to win in life.

I was of the mindset that if I worked hard, the recognition would come. I took a head-down, get-it-done approach, but I was quiet. I hadn't yet learned that squeaky wheels get noticed even if they aren't saying the most enlightening things. Over time, you realize that it's not enough to just work hard. You also have to let someone know what you did, how you did it, and your thinking about doing it that way. For example, being well prepared for a meeting only matters if you speak up and let others know your perspective.

I've been told that I have a presence that gets me noticed, but I had to force myself to contribute during meetings because I didn't always believe my thoughts or questions were that valuable. Taking credit where credit was due or jumping into the spotlight occasionally was never easy for me, and I've learned through talking with other women that it's hard for many—even those with the highest credentials.

At each phase of a woman's career, we experience some common challenges. I've chunked them into the decades where we are most likely to experience them. I acknowledge that these phases may not be the same for everyone, but my experience suggests these will be familiar and hopefully thought-provoking.

LEARNING TO LEARN—TWENTIES

In our twenties, we are learning how to take care of ourselves. This is a time of self-exploration and experimentation. We start with assumptions about ourselves that are informed by culture, upbringing, role models, and training, and we test those assumptions against the situations we encounter. We are learning about having roommates, credit cards, relationships, commitments, and disappointments. It tests friendships as we start to think more independently and finance our own way. I remember those moments of learning to act independently from my family. I recall a time when a family friend had passed away, and I realized I should be writing my own note of condolence to his wife. Our social life is a focal point during this time, and we start to realize the implications of decisions we've already made. Lessons sometimes need to be repeated until we get it right. The wisest women realize how much more we need to learn yet bravely set out to do so.

SocialSelf studied women's struggles with their social life by asking them to rate how motivated they were to improve twenty-one different social life challenges. Midtwenties to midthirties is when women are more willing to "improve self-esteem, shyness, and social anxiety." One key finding suggests that after their midtwenties, "Women are most motivated to become more charismatic."[1] Since early adulthood is when self-awareness increases, and we focus on the traits we wish to develop, we become aware of the need to influence a team, a boss, or a peer to get ahead.

Through our first job interviews, professional roles, and experiences with success or challenges, we also learn a lot about taking initiative and being resilient. By now, we've certainly met our first few go-getters and have been able to compare our ability to influence others with theirs. We've also likely received both positive and negative feedback from supervisors or peers, so we know how we are being perceived by others.

In the *SocialSelf* study, key finding #1 says that women highly rank their interest in cultivating like-minded friends in their early twenties because they want relationships to be more fulfilling. Women also change the way they date for the same reason, relying less on people being physically attractive versus like-minded. That often means looking for partners beyond people in close proximity. Later in our twenties is when we really start focusing on our careers.[2]

During this time in my life, I was traveling Monday through Friday. While working in the cable television industry, I traveled to smaller cities and rural towns as they were the prime locations for new cable television franchises. Once a cable franchise was awarded to or acquired by our company, my role was to hire and train the team of sales and customer service representatives who sold cable packages to customers.

The grueling schedule of travel, not to mention after-hours team-building events, made it difficult for me to maintain personal relationships back home. I remember realizing that I had to work hard to be included in social events with friends when I returned home on the weekends. After a while I was conscious about prioritizing some relationships over others because they were more fulfilling. Given that I had limited time, some friends were not worth the work. I was learning to be strategic about both my social life and my career, aligning with like-minded friends in addition to colleagues.

LEARNING TO ACCOMPLISH—THIRTIES

In our thirties, our strengths are being validated, and we now probably have a handle on our weaknesses. Our reputation is starting to develop, which is the beginning of our personal brand. Traits like being branded a go-getter, thoughtful, or direct are descriptions others have assigned to us, and we are either getting comfortable with those, or we are working to change them. Our values are also playing a role in the things we'd like to change.

We've probably seen enough office and life drama by this point and have begun to align ourselves with issues that are important to us. This is addressed in the *SocialSelf* study, suggesting that women during this phase care less about what others think and more about their own self-worth.[3] Children are likely to be entering the picture, or at least we are considering whether they should be, shifting our priorities and introducing new challenges.

I remember the point after the birth of our second child when I went from driving a sporty car to a minivan after vowing that I would *never* do this. It was a practical identity shift brought on by getting those baby seats in and out of my sporty car. I cared more about making my life easier and was willing to relinquish my sporty image.

During this time, we likely have realized our career trajectory involves some curves and traffic circles. I've often thought it was a cruel reality that a woman's childbearing years were also our prime career-earning years. Sylvia Ann Hewlett writes in *Harvard Business Review* about the realities of career building coinciding with childbearing. "Women pay an even greater price for those long hours because the early years of career building overlap—almost perfectly—with the prime years of childbearing. It's very hard to throttle back during that stage of a career and expect to catch up later." In the moment where we are juggling family and work, it's hard to think longer term.

However, it's important to weigh this: "the persistent wage gap between men and women is due mainly to penalties women incur when they interrupt their careers to have children. An increasingly large part of the wage gap can now be explained by childbearing and child-rearing, permanently depressing their earning power."[4] Women are finding all sorts of ways to address this issue, but our ability to find good childcare is one that changes things dramatically.

Our children were all born between 1994 and 2002. With each child, I took a three-month maternity leave and then did my best to negotiate a return to work that ramped up from three days per week to five. Each time, despite my best efforts, the *off* days became more of a work from home situation. I found that unless I was present to participate in decision-making meetings, I was putting myself or my team at a disadvantage. We didn't have video chat and phone technology at that time, which made participation by conference call challenging. That meant that sometimes it was necessary to physically show up on one of my off days to ensure I was there for decision-making and appeared invested.

Accommodating new mothers was not part of the culture, so if you were in a senior role, you had to be there to defend your position or risk longer-term consequences. Companies are now more open to creative solutions for women because they are thinking more long term too. Replacing valuable female workers is expensive. They know women need to recover, and then upon returning to work, technology supports remote participation in ways that everyone is now more familiar with. Women do have power in these situations, which I did not feel I had during my maternity leave.

Once I returned to work, I was careful to seem as though everything was under control, but it often didn't feel that way. I held a few different roles at AOL since the company was growing so fast, and reorganization to accommodate growth was part of the culture. In each of my roles, I ran large teams with the largest being over 150 people. Almost 100 percent of my counterparts were men who had stay-at-home wives. They rarely mentioned their kids in our small talk exchanges.

I consciously filtered any of my family preoccupations from work chat, sticking to topics they related to. I didn't want to be perceived as unfocused on work priorities or to highlight my other role as a mom. A similarly aged colleague of mine said she didn't put a picture of her kids on her desk for years because she wanted people focused on solely on her contributions, not the fact that she was a mother. We both agree that this was a sad reality of the workplace at that time but also agree it has changed for the better.

I noticed some of my male colleagues at AOL were having career planning/coaching sessions with someone in HR. These conversations included discussions of their career goals and plans that were being put into place to help these goals come to fruition. Those plans included formal executive mentoring, identifying professional development opportunities, and their attendance gave senior executives a chance to get to know them.

It wasn't completely clear how to become eligible for these opportunities, and I remember being disappointed to know I wasn't one of the chosen. Clearly, there was a *list* of people they were grooming for success, and I wanted access to it. I booked some time with my HR representative, but she gave me vague and unsatisfactory answers about how to tap into these opportunities. It was a secret society. I started strategizing about how best to take matters into my own hands.

The initiative taken by a female SVP who ran my business unit struck me as instructional. She made it known to all her direct reports that she was focusing on becoming a Henry Crown Fellow with the Aspen Institute. She tasked us with assembling materials that would support her nomination. We worked to outline significant initiatives that demonstrated her track record of starting things as well as her resilience and grit while benefiting AOL members in ways we could quantify.

They chose each fellowship cohort from a wide pool of accomplished, entrepreneurial leaders who were required to be between thirty and forty-six. If they accepted her nomination, she would be eligible for a two-year series of networking events, seminars, and discussions with other nominees who had "already achieved considerable success in the private sector and are at an inflection point in their lives or careers—looking toward the broader role they might take on in their communities or globally."[5]

This was an opportunity for her to participate in a high-profile learning and networking program that would benefit her for the rest of her career. The idea of seeking corporate sponsorship for such an elite program inspired me because I realized that if she was accepted, AOL would also benefit from what she would learn and the contacts she would make while attending these sessions. It was press worthy, inspiring, and raised her profile inside and outside the company, signaling that AOL-employed executives were participating in global projects. It was win-win.

The SVP was accepted as an Aspen Institute Fellow, class of 2004, and went on to do great things at AOL and beyond. At this time, I had three children at various ages, so my life was a bit of a survival game. But through this experience, I realized that opportunities were not likely to be identified *for* me, so when I was ready, I would have to identify opportunities for *myself*. It should not be left to someone else to determine whether I was worthy.

Career Phases

LEARNING TO NURTURE—FORTIES

During our forties, we likely have a clear handle on what brings us joy and what drains us, but you may not have stopped to consider how these insights should inform your career choices. In a 2018 study by BMC, women identified the most challenging aspects of midlife as "searching for balance in the midst of multiple co-occurring stressors while coping with losses and transitions, for some in a context of limited resources and five themes: 1) changing family relationships, 2) rebalancing work and personal life, 3) rediscovering self, 4) securing enough resources, and 5) coping with multiple co-occurring stressors."[6]

In this phase we are likely juggling the greatest number of balls, which certainly tracks with my personal experience. Steve and I had three children at very different stages of development, parents downsizing and managing illnesses, the transition of a major relocation for our family, day care challenges, and full-time careers. Demands on our time and mind share were at an all-time high, and we were in survival mode. Learning to create boundaries is important at this phase. Sometimes we have to say no.

Not surprisingly, this is a time when women are likely to seek additional guidance from coaches or therapists who support us when pivots may be part of the solution. We are nurturing our children, but we also need help nurturing ourselves, given all that is on our plate. I talked with a few professional coaches for this book and asked for their observations on women at this stage. They confirmed that midlife women who are rethinking priorities are often their primary clients.

Shelly Ryan, founder of Your Next Chapter Coaching, says, "That forty to fifty age range is when women start to really identify who they truly are. I think they've been performing as who they think they were supposed to be up until this point."

Shelly has been coaching women for eighteen years privately, and companies hire her to provide coaching services for their employees as an added benefit. Shelly says that one of the exercises she does with her clients is to ask them to think about what work tasks energize them and which work tasks drain them. Her coaching process then helps them devise a plan to take on more of the energizing activities and less of the draining ones.

Clients then consider solutions such as restructuring their current job, stepping into a completely different role at their company, or shifting their employment status to that of a contractor specializing in their area of strength. In some cases, women choose to take on a new entrepreneurial venture and need help considering all that goes into this big step.

Shelly says that many of the women who come to her are top performers who have hit a stage of burnout. Because these women find themselves to be the *go-to* person in their organization, they have more job responsibilities that don't always come with more pay. They find themselves taking on a greater workload than their lower-performing counterparts, resulting in burnout and hard feelings. Shelly helps them with self-exploration, setting priorities and goals, and then having hard conversations with all those involved as pivots crystallize.

But other facets of women's lives make this time of life challenging. Sylvia Ann Hewlett writes about those who are *high achieving* and *ultra-achieving*. She finds those still childless at midlife wrestle with this reality, too. They have not really made a choice but a "creeping nonchoice." These women find it harder to marry due to a scarcity of candidates. The number of them who were married (57 percent in the US) is drastically lower than their married male counterparts (83 percent of men). [7]

Furthering the comparison, of those high-achieving married men, only 39 percent were married to women with full-time jobs. Only 40 percent of those spouses had jobs making more than $35,000 per year, suggesting a lower level of responsibility at work and, therefore, it is likely that they are taking on more responsibility at home. This is an example of how the same family support systems that are in place for those high-achieving men are simply not there for high-achieving women.[7]

For our family, I remember being exhausted most of the time and fantasizing about taking naps on the weekends. Luckily, I wasn't traveling much professionally. Steve and I worked hard on spending quality time with our kids without trading off the ability to do interesting work. For several reasons, I pivoted into consulting to make this possible, and I write about this in future chapters. Steve and I found it necessary to divide and conquer on the weekends to create these opportunities.

Our oldest was involved in Boy Scouts, where I volunteered, and Steve coached his soccer team. Our middle daughter rode horses, which required time-consuming riding lessons on weekdays and weekend-long competitions, often involving travel. Being more of an animal lover, I took on this duty. Our youngest daughter played competitive basketball through middle and high school, and we traveled all over the US for her tournaments, which greatly influenced summer vacation destinations to fit in both.

All three of our kids also participated in symphonic and marching band competitions requiring parental chaperoning, volunteering, and fund raising. Their foreign language programs also involved hosting exchange students from Japan and Germany for days or weeks. Our social lives became our kids' activities, and that was just fine.

I didn't have the bandwidth for deep reflection about my career until two of our three kids were driving. My focus was on keeping the family functioning and on my own survival. Pivoting into the consulting space provided more flexibility in my schedule. However, I experienced an identity crisis once I was no longer attached to a large, well-known company.

The tradeoff was that I had more control of my time. Where I could, I leveraged professional development experiences that helped expand my contacts and knowledge beyond AOL people. I realized I needed to learn about other types of companies in the DC area because they would become my future clients. It made good strategic sense to start understanding the world of government marketing.

LEARNING TO REINVENT—FIFTIES PLUS OR ANYTIME

This time of life includes a lot of self-reflection. Our self-perception may remain youthful, but how others see us is changing. For example, I was working in the digital media space where the behaviors of younger users of media were completely different from my own. I still read actual newspapers, but only 23 percent of Americans say they still do.[8] As a marketing communications person, I began to feel the need to demonstrate relevance, and I did that by expanding my own social media channels while also aligning myself with young users I could learn from. Any additional time I now have, I invest in trying new behaviors out of my comfort zone.

I also started rethinking the list of things that were unfinished, unsaid, or undone. I call this my UList, but it's commonly known as your *bucket list*. What's on this list will be a direct reflection of your personal brand. I was caught off guard when friends and family started talking about retirement plans because I couldn't imagine ever being idle. But the realities of longtime neighbors moving into over fifty-five communities and friends dealing with health issues or divorces or raising their grandchildren prompted many what-if conversations with Steve. The result is that I've defined retirement as working as much (or as little) as I want, from where I want, on projects that speak to me while still ticking things off my UList.

Kids' ages and phases matter here. Since our kids came later in life, and there is a nine-year age differential from oldest to youngest, we'll be older than most of our friends when our kids reach independence. But given that our kids started leaving the nest when we were in our early fifties, we've had a gradual introduction to this reinvention phase. For others, it happened much more quickly. As Steve and I navigate this, the opportunities seem endless and a bit overwhelming.

Professionally, I know my strengths and have a desire to do things that resonate. I know that companies going through a transition are great places for me to lend my combination of skills. I'm good at building strategic plans around priorities. I'm a consensus and team builder. I'm comfortable operating in a state of flux. I'm good at working at all levels of organizations. What's necessary for me to be driven to work with a company is a mission and vision that I can believe in. I like working with a company and then taking breaks in between to foster passion projects, such as this book! I now have my sights set on lending my expertise to a more strategic role on a corporate board.

Apparently, as people reach retirement age, the choice to stay in the job market is no longer unusual, according to a 2006 Pew Research study. "It doesn't matter if a person is self-employed or not; if a person works for a big organization or a small one; if a person derives a strong sense of identity from work or not; all are equally likely to say they expect to work for pay after they retire. By a two-to-one margin, those who expect to work after retirement say they will do so mostly because they'll want to rather than because they'll have to."[9]

The gig economy makes this more possible, so many people in this age range may need help figuring out how to tap into it. This can be difficult for people with a strictly corporate background and for those who have been out of the job market for a while. For these folks, professional development organizations and strategic networking are a must.

Another factor is that the COVID-19 pandemic changed things for many. After all, this was the pandemic where the risks for older Americans were the greatest. I felt vulnerable for the first time in my life, and that added a sense of urgency to completing the things left on my UList. "Millions of Americans have experienced a tectonic shift in their outlook to work during the pandemic. Americans were raised on the idea of a dream job that could be both personally and financially fulfilling. Work has often fallen short of providing people with what they need to live."[10] For me, turning sixty brought me to writing this book and learning more about publishing. I also started mentoring again, and this time, I'm working with entrepreneurs in a formal way. The goal is to see what new opportunities present themselves. My strategic plan involves using these situations to learn about growing companies so I can strategically lend my skills.

PIVOT LESSONS YOU CAN USE

1. **Find relationships that fulfill.** The social skills of the "Learn" phase teach you the importance of giving and receiving for more meaningful and fulfilling relationships. These skills will last throughout your lifetime.

2. **Source your own opportunities.** You will become more valuable to your current and future employers and clients when you create opportunities to learn, network,

and showcase your work and your company. Look for opportunities that have a high impact and win-win.

3. **Use your power.** All women with new family circumstances have lots to navigate. Have open conversations with your employer to negotiate a reasonable work situation that provides long-term, mutual benefit. Technology is now working in your favor, and it's in your employer's best interest to keep you.

4. **Create boundaries.** Know the difference between becoming valuable and burning out. Say "yes" to the things that matter and create boundaries where needed.

5. **Know what inspires.** Structure your work around the things that energize you and, where possible, shed the things that drain you. Doing so will make you more successful.

6. **UList.** Your UList starts with the things and values that are important to you. Structure your personal strategic plan around it.

7. **Consider a gig.** The gig economy makes it possible to reinvent yourself through projects where you learn and apply new skills. Consider this as an option—especially in "retirement."

Conversations I've had over the years with women around water coolers, conference rooms, playgrounds, sports courts, classrooms, or PTA meetings suggest that at times we are all hanging on by a thread. I've often been hard on myself when family or professional situations didn't run smoothly, despite knowing that perfection isn't realistic and a lack of perfection is not failure. I have also worked to avoid comparisons to others that can create negative loops. Getting comfortable with the cycle of learning, accomplishing, nurturing, and reinventing is another way of acknowledging this idea.

2.

Why Do We Pivot?

Reinvention doesn't only happen after fifty. Sometimes, we find the need to reinvent more than once, and the reasons for doing so vary. There are complete reinventions, and then there are evolutions. In later chapters, I talk about the characteristics, strategies, mentors, and mindsets that support pivots, but here, I want to acknowledge all the reasons we are driven to pivot and then how to move forward.

A PUSH

Others can force pivots on you. This was the case for me early in my cable television career due to a rule that corporate employees needed to rotate into a field office every two years as a way of staying close to customers. There wasn't a formal program to help with these reassignments, but as I approached the two-year mark, I set my sights on offices in the locations most appealing to me: Chicago and Denver. While traveling there on business, I talked with the general managers about what opportunities existed. Being young and single, I would have relocated for the right option, but I wasn't excited about the opportunities available to women with my experience.

Field offices were the places where the company managed local sales and installation of cable. I was interested in expanding upon my marketing skills and became intrigued by a local tech media agency called Redgate Communications. The founder was a charismatic visionary named Ted Leonsis, who dazzled me and all our clients when he talked about the future of tech media. As our primary spokesperson, he advised that businesses experiment with new forms of digital media, such as CD-ROM and online, to inform, engage, and entertain their customers and surpass their competitors.

Ted was a thought leader in the new media space, and he is now chairman and CEO of Monumental Sports and Entertainment, which owns the Washington Capitals, Wizards, and Mystics as well as Capital One Arena. He also cofounded Revolution Growth, a growth capital firm investing selectively in technology-enabled businesses. Ted is a true entrepreneur, risk taker, visionary, and someone I continue to be inspired by from afar. He was always fearless and brazen. The most important thing I learned from him is that if you say something with conviction, people respect it and believe it. For him, this came naturally. For me, this was a developed skill.

This push to leave cable television created an amazing opportunity for me that I probably would not have considered otherwise.

CULTURAL MISFIT

I had a three-month blip between my cable television job and Redgate that I should acknowledge for the clarity it provided me. I joined NEBS for this short time, which specialized in creating paper business forms used by companies of all sizes. NEBS made a competitive offer for the role of maintaining its marketing operations. While the compensation and benefits were attractive, I learned that I needed to be able to innovate and solve problems to be energized by my job. I hoped there would be an opportunity to do that at NEBS, but new ideas weren't well-received, and I felt like a fish out of water. The culture was very conservative and bureaucratic, and I quickly second-guessed my decision. Going to work with knots in my stomach was happening daily. My attempts to muster new enthusiasm weren't working. The idea of restarting the job search felt like admitting failure, but once I asked myself a few hard questions, the decision became obvious.

- How likely is it that this role would change over time? (5 percent).

- Would I be more suited to another role in the company? (They all fell into the same uninspiring bucket for me.)

- What might I learn here? (How to service small a business's administrative needs and how to lower my expectations for myself.)

- Does this company fit with my personal brand and values? (Hard no. Why didn't I see that before?)

Answering these questions helped me see that it was just a mistake. I was excited to learn about the opportunity with Redgate, and the cultural shift was head-spinning and exactly the jolt I needed.

A bonus was that this brief pitstop at NEBS also helped me double-bump my salary, which is something younger generations have now become very good at. Millennial and Gen Z workers change jobs frequently and fearlessly. They have developed resilience due to navigating the chaotic work environment of the COVID-19 recession as well as the recession in 2007 early in their careers. They are now leading the *big quit*. Data from 2023 suggests that 72 percent of Gen Zers and 66 percent of millennials say they are contemplating a career change in the next twelve months with reasons including higher compensation, improved work-life balance, opportunities for career growth, and flexible work arrangements.[1]

This level of comfort in switching jobs was not at all common when I was making my decision to leave NEBS, and my three-month blip was scrutinized during my Redgate job interview. I felt a lot of shame over what I perceived to be bad judgment on my part. Although we never discussed it, my boss at Redgate likely read it as a strength because I had problem-solved and moved on.

MOM-TO-JOB

A 2018 study by *American Sociological Review* looked at whether there were employer biases against those who had opted out of the workforce. The study found that "opt-out applicants were rated lower than unemployed applicants on perceptions of commitment, deservingness, and reliability." It also reports that "opt-out applicants incur similar penalties on perceived capability, which is a measure of perceived skill decline and competence."[2] Women (and men although they are less likely to be in this situation) face "scarring" when reentering the job market after taking a break to care for family members.

Nina Brugel had these very concerns when she considered reentering the workforce. Nina had been home with her kids for twenty years. "I loved that time with them, but I was ready to look farther than my home and family for a greater purpose. In this space of unknown next steps, I felt a lot of disappointment and frustration. I eliminated quite a few jobs because I wasn't interested." As her husband Eric sought acupuncture treatment to manage his pain and stress, Nina began to explore that as a career possibility. She had a treatment, and her interest in the ancient practice grew. But making such a major life change and the pursuit of acupuncture licensing after being a stay-at-home mom for so long was daunting. "Other considerations were the cost of education, the fifty-five miles I traveled to get to classes, and the care of my children since our youngest was still in middle school. I completed three years of full-time classes, logged 250 clinical hours, and sat for three national board exams to finish."

Nina has now been in private practice for ten years and acknowledges that she couldn't have done this without the support of her family. She says, "It is extremely rewarding to help people feel better. I am also so thrilled by the quality of life this job provides since I am in total control of my schedule."

Nina's story highlights two important things. First, she looked for years to find something she was passionate enough about to pursue with this level of gusto. Until she did, she was deeply frustrated but remained open and persistent in her pursuit. Second, she became excited about something that allowed her to maintain flexibility for her family—a common requirement for women who provide care.

According to Dr. Gleb Tsipursky of *Forbes*, "67 percent of women believe that hybrid work has a positive impact on their career growth path. The flexibility of hybrid work allows women to be more efficient and productive, to partake in additional training to support their careers, and to increase their visibility with senior leadership."[3] Nina could think with her heart and head to find an exciting career option. However, not all women have the resources to achieve such long-term goals. These women need flexible options with more immediate benefits.

Flexible and hybrid employment is becoming more of a standard in the current job market, and certification programs are providing new options. According to Bernard Schroeder of *Forbes*, "online and offline certifications deliver incredible returns for the investment, both in time and resources. While a college or university degree takes three to five years to finish, online certification programs run for a few weeks or a couple of months, depending on the field of study." Schroeder goes on to say that "increased employer recognition and the tangible benefits that follow from having a certification to your name have made this a popular choice for professionals around the world."[4]

Indeed provides a list of the most in-demand certifications, and they currently fall into these ten categories: project management, business analyst, supply chain, social media marketing, skilled trade, human resources, sales, accounting, computer network, and health care.[5] It's important to realize that some certifications serve as door-openers while others are necessary hurdles required to improve your career. Analyzing job descriptions of your target companies for preferred versus required credentials will give you valuable insights on ROI before taking on anything very rigorous or expensive.

FAMILY

The COVID-19 pandemic has also caused a shift in how women are navigating the workforce. According to Caroline Castrillon of *Forbes*, "Women aren't leaving the workforce. Instead, they are walking away from their companies in search of better opportunities… they are reevaluating their values and priorities, some are even switching industries or becoming entrepreneurs." Not surprisingly, "almost half of women leaders say flexibility is one of the top three considerations in job mobility."[6]

Because women are also responsible for most of the unpaid work that supports the family, their interest in finding flexibility in their jobs becomes a necessity. "About three-quarters (74 percent) of mothers say they do more to manage their children's schedules and activities than their spouse or partner, and 59 percent say they do more household chores than their spouse or partner."[7] Burnout and mental health have become much more widely discussed since COVID-19, so women have recognized that job flexibility allows them to cover several bases; they can contribute financially to their family, manage the family, and find some space for their own self-care.

In the last year, Monica Pemberton joined the American Council on Education (ACE) as vice president and chief information officer. Her decision to leave a government contracting role she held for over eleven years was based on "my desire to achieve a better work-life balance and prioritize quality time with my children. Although my government contracting role may not have been high-pressure, it demanded a significant amount of my time." When the opportunity for her to work at an association arose, she jumped at it for a couple of reasons. "I recognized that it would provide a more favorable balance, and it represented a promotion in terms of job title. By accepting the new position, I consciously decided to prioritize my family."

Monica was able to foresee that as children get older, their needs evolve, and handling physical and emotional needs are best addressed in the moment. Monica says, "I consider this pivot as one of the best decisions I have made as it has positively impacted both my personal and professional life."

DIVERSITY, EQUITY, AND INCLUSION

Diversity, equity, and inclusion issues are fast becoming part of our political debates and spilling into the corporate environment with large companies starting to weigh in with their voices and through employee benefits and policies. Employees are now influencing their employers by insisting they take a stand when issues around racism, sexism, disabilities, genders, religions, cultures, and sexual orientation are raised.

Take, for example, the issue of CEO Bob Iger speaking out following criticism of his handling of the *Don't Say Gay* bill. Employees felt Iger was slow to comment on the bill. Speaking out would demonstrate support for Walt Disney Company's gay and lesbian workers, and some employees walked off the job in protest.

DEI programs typically "focus on three main areas: training, organizational policies, and practices, as well as organizational culture. Initiatives focusing on policies, practices, and culture exist to correct inequities within an organization."[8] These types of programs also provide a communication channel to senior management when employees feel strongly about an issue and can prevent high-profile protests like the one Disney experienced.

Monica Pemberton also considered DEI and company culture important when taking on her role with ACE given that it was a priority in her prior role with the National Association of College and University Business Officers (NACUBO). She appreciates that she works with a predominantly female workforce and leadership team at ACE. She says there is a "commitment to promoting gender diversity and providing equal opportunities for women in both organizations." Monica is not alone in considering this issue a factor in accepting employment. According to a Pew study in May of 2023, "Six in ten women (61 percent) say focusing on increasing DEI at work is a good thing. But opinions about DEI vary considerably along demographic and political lines"[9] making these sensitive issues a challenge for companies to navigate in the current politically polarized environment.

Given that the gender pay gap has not closed much in the last twenty years, with women currently earning just eighty-two cents for every dollar earned by men,[10] the "E" in DEI efforts will remain an important issue for women as well. Companies that understand the need for a diverse talent pool must prioritize all aspects of DEI to attract top talent and remain competitive.

HEALTH ISSUES

Life sometimes forces unexpected pivots when health issues of our own or others in our family occur. Seema Alexander is founder of Disruptive CEO Advisory and cochair of DC Startup Week. "I had a midlife awakening when I was thirty-six years old when I had my first of two pulmonary embolisms." As the primary breadwinner in her family, Seema was highly successful and working toward being CMO of Prudential Financial, but she was missing something. "I didn't feel like I could be authentically me in a highly regulated environment. I was fighting the bureaucracy of a large financial services company versus implementing transformative strategies that the younger generations were embracing."

Seema was frustrated about regulatory issues and company culture preventing her from building her own brand while she was moving up in corporate America. She had figured out how to play the game, but when her health challenges occurred, she prioritized her desire to run her own business. Having dabbled in consulting on the side, she knew the financial trade-offs, but her ambition and passion moved her forward. Her company has now worked with over one thousand founders who leverage her proprietary UNIQUE Method to transform uncertainty into intentional and sustainable growth.

Seema's story resonates because a health issue in our family pushed me to do some reprioritizing that ultimately set my career on a different course. During the summer between graduating from elementary and entering middle school, our oldest daughter experienced back-to-back illnesses when her doctor prescribed her antibiotics. She suffered severe gastrointestinal pain leading to fear of eating normally, which progressed into a weight-loss spiral. She was hospitalized for almost two weeks, making her unable to start middle school until restoring her weight and energy. This was a scary time requiring specialized care with several different doctors as well as attending partial school days. These school days provided her with a sense of normalcy and allowed her to set and accomplish goals, an important part of her recovery.

At the same time, our son was starting high school, and I was sensitive to not letting our daughter's illness overshadow his big transition. He was our oldest and first child to attend a very large high school, so we were all figuring it out. Since he wasn't yet driving, his activities also required my flexibility. I felt strongly about him building friendships through the new activities now available to him as a way of getting comfortable in this new environment. Steve and I were also suddenly realizing that he would be leaving for college in a few short years.

Both events caused me to quickly shift gears, and I pulled out of the start-up we were incubating so I could focus on these transitions. iBelong had issues I discuss in the next section, but I was no longer invested in trying to resolve them. iBelong was at an inflection point, and so was our family.

As our daughter's health improved and our son found his sea legs in high school, I slowly transitioned into consulting to maintain the flexible schedule I had learned to love. It was so much better for everyone in our family that I could do this. I wasn't necessarily working fewer hours. I just had control over when I was doing it.

These examples of health crises forced pivots in entrepreneurial directions for Seema and me, but we remained in the workforce. Health issues due to COVID-19 have pushed many women out of the workforce. According to 2022 numbers from the US Chamber of Commerce, "women's labor force participation is still a full percentage point lower than it was prepandemic, meaning an estimated one million women are missing from the labor force."[11] Much of this has to do with mental and physical health issues remaining in families since COVID-19. According to the World Health

Organization, "the pandemic has affected the mental health of young people, and they are disproportionally at risk of suicidal and self-harming behaviors. Women have been more severely impacted than men."[12]

LEADERSHIP OPPORTUNITIES

"Decades of studies show women leaders help increase productivity, enhance collaboration, inspire organizational dedication, and improve fairness. Despite these benefits, only 10 percent of Fortune 500 companies are led by women."[13] This reality, combined with the knowledge that "female employees are less likely to be promoted than their male counterparts, despite outperforming them," cause women to create their own opportunities for leadership. More frustrating is that in that same study, "women received higher performance ratings than male employees, but received 8.3 percent lower ratings for potential than men."[14] This study, as well as women's own recognition of the gaps in performance and opportunity mean women are finding other ways to get ahead.

Part of the reason I invested in iBelong was because it was a logical next step given my entrepreneurial background, but the other was that I was interested in becoming CEO. I knew I had some things to learn, but given my cofounder's background and network, I felt I would have that guidance. I was also aware when starting a new company, an organizational chart is meaningless until you build a company worth running. I expected that investor contributions of time, money, ideas, and talent mattered to assigning formal roles. I had not considered that money from other investors could undermine this assumption.

When a new investor made his investment contingent on holding the role of CEO, it took me by complete surprise. I thought I understood the playing field, but clearly, I didn't. I tended to my hard feelings but worked to support him because, after all, I had an investment to protect. Watching the company start to unravel from some of his management decisions was difficult. My daughter's health issues coincided with this unraveling, and I lost the desire to keep trying to turn things around. I cut my losses and left the company.

This powerful lesson took months to digest. I got firsthand experience in start-up financing and the power dynamics associated with it. I write more about this in chapter 5, "Entrepreneurialism," and this experience contributed to my interest in making sure women have opportunities in the C-suite and in the board room. While companies going through growth transitions continue to be a sweet spot for me, I now have a career's worth of experience to lend when I take on fractional CMO roles for my clients.

MERIT

Janet Hall is currently the founder and principal of the Cortical Group. After finishing her MA in Political science, her first career stop was working in a government agency. She says, "I really liked what I was doing. It was fascinating, but I also was bothered by the bureaucracy. Yet I understood that bureaucratic constraints were there for good reason. I had the opportunity very early in my career to go overseas for a plum assignment and was doing very well. I was in Germany during a hugely exciting time when the Berlin Wall fell, when Eastern Europe fell, and the Soviet Union in a coup." Janet realized afterward that she had just been up close and personal with some of the most exciting things she would likely see in her career. "Returning from these assignments, you still get promoted, but they begin to slow you down. I was curious about what else was out there. Specifically, I wanted to test myself to see if I could produce the kind of results that came with generous compensation."

Janet made her first career pivot when she decided to go back to school full time to get an MBA. She says, "I didn't really know how to find a job in the private sector or how to talk about myself. In hindsight, I could have done a lot of different jobs with the background I had, but as a female, I felt like I needed more education. The MBA was good for giving me the vocabulary I needed for business."

Janet was fortunate that her MBA was through a fellowship at the University of Maryland, so she got a free ride, but she says she probably didn't need it to get her next job. She wasn't sure how to start making the transition to the private sector, so she sought the guidance and security of an MBA. She found the business vocabulary she acquired to be very useful.

Feeling the need to be exceptionally qualified for roles is not unusual for women. A *Harvard Business School* article suggests that women should consider different strategies to be successful in the workplace than they used when being successful in school. We are used to great effort being rewarded with high grades in school. For women in the workplace to get great rewards, they need to shift their thinking. "To be successful, we must now do the very thing we were always taught not to—be disruptive."[15]

Two noteworthy concepts are raised in this HBR article. First was the idea of finding effective forms of self-promotion. Doing things like cultivating *and* mobilizing relationships on a regular—not just occasional basis—is uncharacteristic for women but very necessary to getting ahead. Speaking for myself, I'm good at cultivating relationships but not as good at mobilization. The article suggests that you should be constantly and strategically putting relationships to use. The ways you choose to self-promote will have to be more subtle than men to strike a socially acceptable tone.[16]

The second noteworthy concept raised was the idea of choosing a less prescribed career path. "More and more women are embracing unusual, self-directed career paths that play to their strengths and are aligned with their values. When you are scared, consider that to be a good sign."[17]

RETIREMENT (REINVENTION)

The average age that a woman retires is now sixty-two.[18] I'm at the age where people all around me are retiring, and I'm fascinated with the different ways people are choosing to do it. I'm seeing big trips to exotic locations being planned regularly and downsizing into fifty-five and over communities for less maintenance and easy access to social activities. I'm also seeing experiments in rural living, home flipping, second careers in teaching or mentoring, community service, advisory services, small business creation, issue-oriented work, board work, and living out fantasies to putter. For me, looking back was essential to moving forward. I had to do some analysis, some shedding, and then some dreaming and planning. This book was a big part of that. I consider it my sabbatical before architecting the next phase.

Women have the opportunity to get creative in retirement, but it's worth mentioning some unique challenges they have in financing retirement. "About 50 percent of women ages fifty-five to sixty-six have no personal retirement savings, compared to 47 percent of men." A woman's current or former marital status also impacts how much retirement savings she is likely to have, as does the presence of children and with how many different partners.[19] Given that a higher percentage of women may need to provide for themselves in retirement, their prior work experience and ability to reinvent themselves are relevant.

Kristin Runke has been a self-employed optometrist for thirty years and is now trying to decide whether and how to retire from her beloved practice. She is thinking beyond the financial aspects of retirement as she makes sure she and her husband can consider other things. "I love the job, yet I feel the lure of being free to travel and explore painting as well as focusing on fitness. My eighty-six-year-old mother lives with us, and I know I will also be pivoting as a caretaker in future years. Life has stages, which I must remind myself to enjoy."

I appreciate Kristin's sentiment of reminding herself to enjoy the options that this stage of life brings. I'm guilty of not always taking the time to smell the roses. But I'm also very aware of not revving down my ambition engine just because I've reached the age where some do. Retirement is not one thing but a balance that allows us to take advantage of the opportunities we now have.

PIVOT LESSONS YOU CAN USE

1. **Pushes provide lessons.** A push will bring learning if we take the time to notice. It allows us to rule in and rule out options.

2. **If the shoe doesn't fit.** You will likely regret joining an organization or taking on a role at some point during your career. Many of us do. Don't stay in situations that aren't right for you.

3. **Keep looking for passion.** Finding your passion takes time, and it may come after raising families. Look for the inspiration in your daily life and then chip away.

4. **Company culture should reflect your values.** Test for the amount of diversity, equity, and inclusion that suits your needs and values. Companies that are DEI savvy are fast becoming the ones younger workers seek.

5. **Flexibility is a common priority.** Career choices aren't binary, and playing to your circumstances is sometimes necessary.

6. **Curveballs get thrown.** Unexpected health issues in families commonly affect women in the short and long term. When health pivots become necessary, patience becomes a virtue.

7. **Seek leadership.** Leadership opportunities aren't likely to present themselves to you. You will have to go seek them out.

8. **Be disruptive.** This goes against everything we were raised to be, but try being disruptive at least once a week.

9. **Mobilize relationships.** Merit is only half of the equation. Cultivate and mobilize relationships while self-promoting in a way that is authentically subtle to you.

10. **Reinvent for retirement.** Everyone's retirement circumstances will be different, but the ability to reinvent ourselves at any age is important.

While these don't *represent all* the reasons why women pivot, they represent the ones I've encountered personally and those raised by women I've talked to. Regardless of whether the list is exhaustive, you can apply the lessons provided to more than one situation and speak to the unique circumstances we continuously find ourselves in.

3.

Navigating Organizational Chaos

Size may matter. During my career, I've held roles with many different size companies, ranging from a small marketing communications firm yielding $13 million in revenue to a large online media company with $600 million in revenue and various sizes of companies in between. In my consulting business, I've also worked with nonprofits of all sizes, who rely on grants and donors as well as government contracting firms that service the federal government. I cofounded a marketing software company that raised venture capital, but we had management and technology issues that we didn't survive. The common thread for all these roles in for-profit, not-for-profit, government services, and tech has always been my interest in building something. Being close to the visionary was important. I never lasted in roles where the road map was already in place. Building the company helped me become invested in the mission of the organization.

CHARACTERISTICS OF ORGANIZATIONS

Corporate organizational structures are reflective of the size of an organization and the way decisions are made. For example, the amount of revenue usually dictates how deep the management hierarchy runs, as does the complexity of the market/product/sales cycle and the skill levels of the employees and managers who work there. As tech firms came on the scene in the mid-nineties, they realized that their work should be done differently, which impacted hierarchies, operational structures, and pace of work. Communications became digital and faster, and decisions started being made through email. Work became more twenty-four-seven.

In larger organizations, priorities are often set several levels above most employees, and unless internal communication is thorough, it can be difficult to follow along. In these organizations, roles tend to be more specialized because of the larger volumes of each type of work. It may also take longer to see bigger projects through because multiple departments and larger teams have to be considered.

In smaller organizations, you are more likely to wear many hats, making roles simultaneously broad and deep. You might find yourself doing your job while supervising the work of others or managing a department with a broad function. Because employees are closer to decision-makers, they are likely more aware of events that could shift priorities (e.g., negative feedback from an important customer or employee), and therefore, a characteristic of survival may be adaptability.

Every organization has its own personality and culture, yet understanding typically common characteristics driven by industry and size helps you find one where you'll thrive. These small/large company differences became apparent to me when our smaller company was acquired in 1996.

Redgate Communications was a scrappy *new media* marketing firm that helped our clients navigate the new digital landscape. Our CEO, Ted Leonsis, was the visionary who saw the future of consuming entertainment, information, and advertising messages. Ted met AOL's CEO, Steve Case, as he was pioneering the use of his new consumer software product that helped customers navigate digital information. Together, they outlined a path for bringing Redgate clients' content onto the AOL platform. AOL subscribers would get exclusive access to this content in new, interesting, and graphical ways. For example, Fidelity Investments published investment fund information on AOL. Marriott Hotels published information about its property locations and amenities.

When our company of less than one hundred employees became part of AOL's new growing tech media company, navigating this environment required a different approach. I took note of a few new realities. The first was that mergers are chaotic. The second was that company culture encouraged and accepted irreverent behavior. Everyone had to figure out how to thrive and survive. If you aspired to leadership, you also needed to find ways to stand out in this wild, new environment. Finally, you had to be devotedly agile. Emails at 2 a.m. were not uncommon, and the expectation was that you were always up to speed on the current thinking of whoever was in your email inbox.

ACQUISITIONS AND RELOCATIONS

Leading up to the acquisition, we Redgaters discovered that moving to Northern Virginia would be a condition of our employment since AOL headquarters was there. Steve and I had just become new parents, and adding in a relocation at that time seemed overwhelming. Each Redgate VP had a predestined role within the Interactive Marketing Group (IMG). The head of that group scheduled interviews, and many VPs willingly started the relocation process.

I was less enthusiastic but was operating from a position of strength. For several months before the merger, AOL functioned as a client of Redgate. I was the VP servicing the AOL account and regularly flew from Boston to Virginia to meet with AOL division heads. As I sourced and managed new business with my AOL clients, I got to know many department heads, identifying whom I would eventually like to work for. I was evaluating leadership personalities and business opportunities with an eye toward staying in Boston if possible. I was able to find something that felt like a perfect fit.

After the acquisition became official and during my IMG interview, I let that department head know about my preference to take a role that didn't require my relocation to Virginia. I planned to accept an offer to launch AOL's Digital City Boston, which was to be the local online city guide for the Boston market. We would compete directly with *The Boston Globe*'s new online product called boston.com. Digital City's mission as a start-up inside AOL was to create a presence in all major US markets with Boston being a priority city. Me taking the role as general manager, or mayor as we were titled, would move Boston to the top of the launch list.

My plan to accept this role was not well-received by the IMG department head I was interviewing with. After some screaming and desk slamming, he stormed out of the room, muttering, "We'll see about that." No doubt, someone in HR got an ear full because these placement decisions were being made for us, and I was undermining the process. That encounter rattled me, but shaking off this type of aggressive behavior was essential to survival at AOL. By comparison to other outbursts I would later witness, this one was pretty mild. When in situations where behavior was similarly aggressive, my strategy was to be quietly unemotional and to let it burn out.

In the end, I did stay in Massachusetts, running Digital City Boston for five years. While I didn't have to relocate initially, there was regular travel to the headquarters' office for meetings with other mayors and Digital City's corporate team. It was still less travel than I would have done in the IMG because that group was constantly pitching advertising clients across the US.

Remaining in Boston also meant it didn't disrupt Steve's job at a critical time in his career. We didn't need any further chaos as new parents. We were still figuring out the basics of feeding schedules, getting enough rest, managing childhood illnesses, day care, and getting out the door in the morning. After three years, we felt like we had mastered team parenting, and we added our second child into the family.

NEGOTIATING MATERNITY LEAVE

Announcing and navigating my second pregnancy was stressful and challenging, requiring patience and self-advocacy as well as a leap of faith. Given that the US is still the only country with no national paid maternity leave, we can see that not much has changed toward making women feel less vulnerable.[1] Since I was in a field office in my role with Digital City, my pregnancy was not obvious for a while. I announced my pregnancy around five months in a face-to-face meeting with my boss.

AOL's maternity policy guaranteed twelve weeks of leave (not all paid), and the terms of your leave and the job you returned to were based on a negotiation with your supervisor. Since we were a start-up within AOL, it was never clear how separate and, therefore, different our benefits were. Digital City Boston had only been live on the AOL service for about six months when I announced my pregnancy, so we were still building that business.

Given that my boss was also juggling the launch of several other Digital City markets, it was difficult to get his attention on my temporary succession and reentry plan. I tried forcing conversations by purposely standing in profile in team photos for the company newsletter so he could see that my pregnancy was progressing. At the eleventh hour, we finally agreed on a play-it-by-ear strategy. I could come back to a job at AOL, but it wasn't clear which one. Despite my best attempts to get something in writing for my own peace of mind, we left things open-ended. I tried to stay engaged to avoid any surprises upon my return, but I was determined to have this time at home with our new baby.

After twelve weeks of adapting to life as a family of four, I did return to my role as mayor, but the pressure to move to Virginia was ramping up. A legitimate business case could be made for opening a local office in a new market, but eventually, business operations needed to be centralized, especially when Tribune Media became a new investor in the venture. In 2000, when our kids were six and three, our family answered the call to move to Virginia. It was becoming more obvious that opportunities were going to diminish unless we took this step. We AOLers believed we were changing the world, and I wasn't ready to step away just yet. Steve and I rationalized it was best to do this while our kids were young so the transition would be easier for them, but it's hard for your kids at any age. Steve became a stay-at-home-dad for our first year in Virginia, and it made that transition so much smoother.

During this exciting and chaotic time in our family, Steve and I worked to manage ambition with our desire to have a *normal* family life. For us, normal meant finding ways to connect with our kids every day and making sure that our jobs didn't prevent them from having access to opportunities, activities, and friendships. Providing this access became more challenging as they got older, but with our younger kids, we could manage.

Initially our son experienced some adjustments entering a much bigger school than he had left behind. Our daughter started preschool in Virginia, and we learned how challenging preschool access can be. There were applications and interviews, believe it or not. Ultimately, Steve was able to restart his career, where he left off.

With few exceptions, my colleagues who relocated to Northern Virginia were either unmarried women or men with stay-at-home wives. Waiting five years to make this move was essential for our family. It may have been professionally risky to do so, but I mitigated the risk by being flexible in other ways. Regular appearances at the corporate office so that my presence was always felt was one way. Since I lost a senior-level advocate early in the acquisition by not joining the IMG team, I worked hard to find others. When Steve and I finally did move, we had a sound plan and were ready for this next adventure.

SURVIVING REORGANIZATIONS

During the ten years I was with AOL, I lived through three significant reorganizations. These were not unusual because the company was growing so fast. Janet Hall worked in the AOL Product Marketing Group, but we didn't meet until after we both left AOL. One of her first roles at the company was working on the AIM product, the precursor to the instant messaging products we all use today.

Janet has since gone on to senior-level marketing roles at FiscalNote and the American Heart Association and now runs her own consulting practice. When she and I compare notes about working at AOL, we often talk about how formative it was and about the many strong personalities we encountered there. In interviewing her for this book, she said, "It was a crash course in entrepreneurialism, even though it was already a public company. It was also a crash course in chaos."

For Janet, that crash course started when she first interviewed there. "AOL was doing a bunch of renovation, so when I came in for the interview, we had to sit on chairs in the foyer because they were hammering and putting up cubicles all around us. People are coming and going, and we could barely hear each other." The company was growing so fast that there was a frequent need to add more cubicles, floors, and, eventually, buildings.

Only three months after she joined, a change in AOL's pricing made the product she was hired to work on extraneous. She would need to find another job inside the company, but as was typical, she was on her own to do so. She did what we all did during these reorgs. We worked the list of open positions and got in front of the hiring managers.

Janet says she also considered looking for another job outside of AOL because these reorganizations were unsettling. "I'm glad I didn't find one because I ended up being very happy there running the AIM software division. I learned a lot about people and managing uncertainty while I was there. That has been hugely useful to me in my career, no matter the size of the organization I've worked in since."

Janet agrees that as AOLers, we felt like we were changing the world, but it's hard work operating in uncharted territory every day. These circumstances made maintaining self-confidence a challenge. The situation required boldness, agility, and resilience, but that can be hard to muster if the direction changes frequently. In this entrepreneurial culture, taking risks was encouraged. However, a lawless environment added a level of complexity to our roles as managers of teams.

Bad rumors could cause consequences if employees felt at risk. For example, speculation about products being sold off or another reorganization coming down the pike might tempt valuable employees to look for new roles that felt safer before an undesirable option was dictated to them. The active rumor mill contributed to the chaos and added another dimension to the chess we were all playing. We had to be on alert, protect our teams, keep our agenda at the forefront, and quickly calculate our next move as changes occurred. This all provided a great lesson in corporate survival.

Nancie Laird Young, another former colleague from AOL, spoke to me about her experience with the company and the challenges she had navigating that environment. Because of her personal circumstances, she said, "My therapist told me not to take the job at AOL. She said please do not go there. I have clients from there, and it's the most toxic environment." Nancie did take the job. Her background as a community manager with washingtonpost.com made her the perfect person to supervise the discourse between members in various communities on the AOL service.

As Director of AOL Community, Nancie supervised the team that moderated online communities organized by special interests (pets, plants, social groups, geography, religion, political affiliation, fan clubs, kids, you name it). In these communities, AOL members would share advice and ideas, but when conflicts arose, safety needed to be considered—especially for kids. If those conflicts escalated, a moderator would step in.

As part of her role, Nancie was the ultimate escalation point for these conflicts. She played mediator and had the control to take away member privileges if they violated the AOL Terms of Use. This was the very early stages of the content moderation issues we are now seeing on modern social media platforms, and again, we were breaking ground.

Nancie's was a stressful job, and she now says this about arriving at AOL: "I broke a rule I'd lived by my whole life; take time to learn the geography, the lay of the land, the culture." Because Nancie was joining some former colleagues from washingtonpost.com when she arrived at AOL, she was expecting the culture to be familiar. "I thought because I was working with some people who knew me, I would slip right in. But since I also worked with many people who did not know me before, it took years to prove myself in their eyes."

Nancie makes two great points. The first is that personal circumstances must be considered when taking on a new role. AOL's twenty-four-seven culture, combined with the type of high-stress job she took on, along with her primary role as a single mother of three, made her move to AOL extra intense. The second point is that even though colleagues may be familiar, each organization has its own culture, and it should be carefully researched (as I found out with NEBS). When making decisions about joining organizations, "77 percent of job seekers take company culture into consideration." But given that it's so intangible, it may be hard to measure. Simply asking about the company's mission, vision, and values probably provides great clues.[2]

Nancie goes on to say, "Constant reorganizations made it difficult for us to do our jobs because new management came in with different priorities, not understanding how online communities work." I was one of the new management folks Nancie is referring to. My charter when I started managing this group was to find ways of monetizing the online communities. That meant trying new things on behalf of paying advertisers.

As the one charged with disrupting the peace, I was the catalyst of Nancie's angst given her charge of keeping the peace. In frank conversations, we've both acknowledged how crazy it was to navigate the environment at AOL, but she says, "I found you so honest and accessible, and I trusted you. I always felt you had my back." It means a lot to know this was her takeaway.

PIVOT LESSONS YOU CAN USE

1. **Pick a size**. Understand the characteristic differences between small and large organizations and which fits best with your appetite for adventure, stability, or influence.

2. **Merger prep**. Acquisitions and mergers are turbulent. They just are. Don't wait for the organization to figure out all the answers. Your own research will matter to influence a win-win option.

3. **Maternity leave is important.** Don't shortchange time with your new baby. Everyone involved benefits.

4. **Measure risk tolerance.** What's going on in our personal lives must be considered when we take on a new role. Set yourself up for success by making sure the time is right.

5. **Research company culture.** Do your research and listen to your "spidey" sense on whether a company's culture matches your temperament and goals.

6. **Changing the world is hard work.** Doing groundbreaking work requires a combination of fearlessness and compassion. Make sure you are ready to handle the turmoil that comes with it.

7. **Mitigating chaos.** This requires compassion, an interest in adventure, agility, and resilience.

AOL was formative for me at a time in my career when our family was growing, but my leadership style was also developing. The people who tended to succeed there were often brilliant, brash rule-breakers who could outmaneuver their counterparts and boldly push their agenda. Performing your job well was only part of the equation. Adaptability and political savvy were essential survival skills I had undervalued until working there.

4.

Identity

———

"The trouble with you is that your expectations are too high!" my mother said to me. I had just told her that my date the night before was a perfectly nice person, but because we looked at the world differently, I wasn't planning to see him again. But my mother—concerned about me being in my late twenties and unmarried—thought I was too choosey and opinionated—not characteristics that women in my family were encouraged to embody. My maternal grandmother was forceful and opinionated in spades, and I so admired her for it!

Developing a strong self-identity comes with the experiences of making your own decisions and being right some of the time yet surviving the consequences when you are wrong those other times. But to have those experiences, women must be willing to put themselves out there and follow their instincts. One of the reasons I started writing this book was because I was curious about why some women are willing to do that more than others. Is it personality, upbringing, geography, parental expectations, opportunity, or something else? According to the cross section of women I spoke with for this book, the main common denominator was an intense curiosity that seemed to outweigh whatever fear they might experience. They all had varying degrees of opportunity. Some were more encouraged by family members than others, and they grew up all over the globe, but their curiosity drove them. Fearlessness only developed with experience. Many say they still don't feel fearless but are curious to see what they can achieve.

SELF-ESTEEM AND MOTIVATION

In talking with women about the idea of how their self-identity evolved and whether that inspired a pivot, conversations often focused on the fact that careers are not the linear hockey-stick-shaped, upward trajectory we often thought they would be. For women interested in raising families, continuous child-rearing causes us to reevaluate whether the work we are doing is worth the trade-offs we find ourselves making.

The shift from professional to mother/professional can bring on an identity crisis, which sometimes challenges us to pivot. For those women not raising children, the investments they have made in their educations and careers have shaped their identities. The idea of making changes to this are difficult to explain and potentially causes financial instability.

According to an article summarizing a few different self-esteem studies, authors Stets and Burke identified three different motivational aspects that impact self-esteem. Once motivation is realized, action is likely to follow. The first motivation outlined is the self-enhancement motive, driven by people's need to belong and be liked by a community. The second is the "agency motive," reflecting someone's ability to "affect the environment" and a desire to be "in control of the forces that affect one's life." The third motivation is the authenticity motive, which deals with "meaning, coherence, and understanding about the self."[1]

In the stories below, we will look for which of these motivations may have been at work, but taking action may require something more. Your ability to act is related to inner strength developed through experiencing success and failure during your lifetime. Michelle Icard's book titled *8 Setbacks That Can Make a Child a Success* talked about how failures our children experience contribute to making them confident adults.

As parents, we try to protect our kids from adversity, but Icard says this is doing them a disservice. They must understand resilience to deal with what life will bring their way. Although written as a resource with amazing parenting tips, the idea of failure breeding trust and confidence in our decision-making ability is relevant to having the courage to pivot. "When you learn that even if the world around you can be unpredictable and disruptive, you can ask yourself, but do I trust myself, and if you can say yes, that is based on experience. And it's really what sets us free."[2]

In the following stories from Irene, Jenna, Elaine, and Jen we can identify motivations, but we should also acknowledge their confidence to refine their interests and push for greater clarity to pivot.

ARCHITECTURE → SOFTWARE ENGINEER

Irene Hakes has an undergraduate degree in architecture. After seven years working for two different architectural firms, she started interviewing to find a better company match. She finally realized the nature of the work, not the firm, was making her unsatisfied. Given her love of solving puzzles, and with the encouragement of her husband, she started taking coding tutorials and quickly realized she was good at it.

She now has a successful career in software engineering. She says, "The biggest challenge for me was the initial thought that pivoting somehow meant I had failed at my goal to be an architect. In reality, I had achieved that, but there was still a sense that because I was 'giving up' on that as a full-time career, I had wasted my time."

She goes on to say that "the fear of having 'wasted' my time proved to be misplaced. Quite the opposite. The professional skills I learned while working as an architect have proven to be a huge asset." Irene says several of her friends from architecture school have also transitioned to a different career at least once and some multiple times, "and it was always because they needed better work/life balance." Irene now describes herself as an "architecturally trained software engineer."

When Irene interviewed for her first software engineer position, she had to face some questions asking her to defend her reasons for making the change. She said questions often varied: "What will you do if you no longer like software engineering? How do you know this is what you want to do for the long term? The truthful answer was always I don't know, but for now, this is what I'm committed to."

Irene trusted her instincts, and we can see her authenticity motivation at work. She leveraged her sophisticated creative and problem solving skills to design a new career requiring many of those same skills. Knowing Irene, she came across as genuine, thoughtful, and amenable, so convincing a prospective employer to take a chance on her was not a stretch.

Realizing that other pivots may not be as seamless, it's wise to take a lesson from Irene's approach. Being honest about your reasons for pivoting, drawing any parallels between the new and old roles, demonstrating a willingness to learn, and a commitment to the task are all winning strategies.

ACTING → VALUES-BASED VIDEOGRAPHY

Jenna Close is the co-owner and director of photography for Buck the Cubicle. Her company does storytelling through video and photography for purpose-driven brands that want to give voice to their values. This interesting niche of showcasing a company's values in creative ways took some time for her and her colleagues to land on.

As a corporate and industrial photographer for years, Jenna says she was starting to feel stale. "Buck the Cubicle started as a personal project. To try to reinvigorate our creativity, we decided to do this whole series of videos on people who were really passionate about odd jobs or hobbies. It was so much fun and so well-received that we ended up just transitioning our entire business." This experience reinforced that she likes working with companies doing something "beyond just making a profit and wrecking the earth."

To get here, Jenna pivoted a couple of times. When she was in her early twenties, she was fulfilling her dream of becoming an actor, but suddenly it wasn't fun anymore. "I fought that instinct for a long time because I had such a rigid sense of who I was. This is what I'm supposed to be doing, and I've worked so hard to get here, but I finally just had to walk away." Jenna says her whole identity was wrapped up in being an actor, and because it was so painful to transition, she did it very abruptly and damaged some relationships in the process.

Eventually, she was able to combine her creativity and visual media skills with her interest in responsible and ethical brands. Her decision to invest in a "personal project" sparked creativity, and the recognition she later received propelled her to keep going. Now she asks her prospective clients some key questions to make sure they meet Buck the Cubicle's criteria for supporting brands that have an ethical message. Jenna now describes herself as someone who "enjoys telling other people's stories, meeting people around the world, and showcasing the best of humanity."

Jenna's pivot was years in the making. She kept pushing the creative envelope, and these instincts, along with her authenticity motivation, defined her business strategy.

GOVERNMENT → MOM → ENVIRONMENTAL EDUCATION LEADER

Elaine Tholen spent twelve years doing environmental cleanup work for the Department of Energy before deciding to pivot home full time with her two sons. While at home with her family, she volunteered at the National Wildlife Foundation, where she led programs for kids. She also participated in a Master Gardener's Program and began volunteering with schools to develop schoolyard gardens and wildlife habitats. Eventually, she became active at the state level, promoting environmental education.

After a quick return to the Department of Energy, she validated that her heart was not in this higher-paying work, but her "true passion was the intersection between K–12 education and the environment." Once her children were old enough and childcare was no longer necessary, she expanded her networking and volunteering, which evolved into a few different part-time jobs.

Elaine says, "After working like this for about two years, our program had grown from six schools to over fifty schools, and more were asking to be involved. The school district decided to hire a full-time person, and after many interviews, I received the dream job I had created." Elaine says there were monetary challenges to this decision and, of course, a time commitment. "I was moving from a fairly lucrative career path to one that would be hard earned with little monetary gain." Elaine now describes herself as an "environmental education leader and advocate for student engagement that thrives on the energy of young people."

Elaine credits her "years of networking with organizations all over the state and the Washington DC area" with her ultimate success in fully pivoting. During that time, "I built strong partnerships with organizations to provide teacher professional development, student programs, technical assistance, and awards and recognition for the work we were doing. The Get2Green program I developed is nationally recognized, and Centreville Elementary School is nationally recognized as one of the greenest schools in the US."

Elaine's story taps into both the agency motivation, where she could control her schedule for her family, and the authenticity motivation because she tied together her interest in the environment and in kids. Also notable is that her plan took ten years to ultimately realize.

WHITE HOUSE → JOURNALIST

Jen Psaki is a role model of mine. I admire her grace under pressure and experience working at the highest levels of communication and government. After doing campaign work with Tom Harkin for the US Senate, Tom Vilsack for Governor of Iowa, and John Kerry in his 2004 presidential campaign, she traveled with Obama during his 2008 presidential campaign as his Press Secretary. After Obama's election win, she followed him to the White House. In 2015, she became Obama's Communications Director. She also held senior communications roles with the US State Department and for Global Strategy Group. She was named White House Press Secretary for the Biden administration. She is now a correspondent for MSNBC.

Jen has talked about how her tone was so important when interacting with the press while she was in the White House—not just because she is a woman, but because tone needs to match the moment to be an effective spokesperson. Despite her credentials, she has confessed to suffering from impostor syndrome.

"I have had moments in my career when I know I can do it, and I have the experience to do it, but you have doubts. And part of it is using that not as a weakness but as a source of motivation and strength."[3] Jen says the best career advice she's ever received came from former White House Press Secretary Robert Gibbs, who told her, "If you act like you belong there, at a certain point, people will believe you."[4]

As an early target of the Russian propaganda machine and while she was new to her role in the State Department, public attacks on her started becoming very personal. "They talked about what I wore, how I look. But at the core of their message was that I had no idea what I was talking about and that I was dumb. They were hitting at something I was already sensitive about. That is their playbook. When this was all happening to me, I really didn't know what to do."[5] A colleague told her to consider this a badge of honor that she was being targeted by the Russians because that meant they considered her a threat. This perspective helped her to find the strength and motivation to push on with her work.

Jen's story is a very high-profile example of someone wrestling with self-esteem under high-pressure circumstances as she worked hard to quell the spread of misinformation. She performed her job so well that we never saw her flinch. I'm reassured to know that someone of her caliber has moments of self-doubt. Given her willingness to speak openly about what she was feeling behind the scenes in the Obama administration makes me even more of a Jen Psaki fan.

When she pivoted in March of 2023 to her current role as host of "Inside with Jen Psaki" on MSNBC, she faced criticism again. In an interview with Kara Swisher, Jen was chastised for taking six months to get her new show off the ground and then for her early guests being all Democrats. Jen responded by saying, "People have to agree to come on the show, of course. I do want to have Republicans on the show. I've only had four shows."[6] In the early stages of this role, she was learning on the job yet taking criticism again and was unflappable. Jen is clearly talented, and I very much appreciate her relatable honesty.

Jen is clearly familiar with all three self-esteem motivations, but more importantly, she has experienced several confidence-building setbacks that she has grown from. I'm sure her openness is appreciated by many women besides me.

PIVOT LESSONS YOU CAN USE

1. **Be curious**. Approaching challenging situations with curiosity versus trepidation makes it more likely that you will be able to push through difficulties. A curious state of mind takes pressure off the outcome and allows you to focus on the process.

2. **Examine motivations.** As humans, our motivations are quite simple. We want to belong, we hope to exercise some control in situations, and we look for meaning that is important to us. Examining our relationship to each of these core motivations enables acceptance so you can move toward an outcome.

3. **Build confidence and trust.** Going through experiences with challenging outcomes builds trust in your ability to manage adversity. We are braver after testing our own survival skills.

4. **Pivots take time.** It could take years to navigate a pivot, making trade-offs along the way. Keep your eye on the end goal.

5. **Act like you belong.** Acting like you belong means showing up, dressing for the role, and acting the part. Others can't believe you unless you believe yourself, so lay the groundwork for success. Whatever it takes.

6. **Dispel doubters.** When making a pivot, doubters may challenge your decisions and ability. Authenticity helps. Match tone to the moment.

Practicality is important when pivoting, but so is listening to the motivations driving you from the inside. Irene, Jenna, and Elaine have each talked about how they now self-identify. This has become part of their personal brand, which I write more about in chapter 7: "Unexpected Ways of Impacting Personal Brand."

5.

Entrepreneurialism

My Catholic upbringing suppressed my ability to break rules without conjuring massive amounts of guilt. Despite this, I thrive in entrepreneurial environments. Entrepreneurs are known to be unbound by conventional approaches to solving problems, and I appreciate those who think in different ways. As she wrote in Harvard Business School Online, Kelsey Miller lists these as the characteristics of successful entrepreneurs: "curiosity, structured experimentation, adaptability, decisiveness, team building, risk tolerance, comfort with failure, persistence, innovation, and long-term focus." [1]

Given that I recognize many of these characteristics in myself and the entrepreneurs I've known, I felt it was a good frame of reference for those considering entrepreneurship. Gauging one's own ability to thrive in these environments requires honesty about your strengths, weaknesses, and personality. If predictability or a clear road map is important, you may want to reconsider. Managing uncertainty, thriving during experimentation, and being resilient will be necessary for survival in true entrepreneurial environments.

TYPES OF ENTREPRENEURS AND PERSONALITIES

In the book *Built for Growth: How Builder Personality Shapes your Business, Your Team and Your Ability to Win*, authors Kuenne and Danner "discuss four distinct types of highly successful entrepreneur personalities—the driver, the explorer, the crusader, and the captain. Each is motivated, makes decisions, manages, and leads their businesses differently."[2]

I see myself as an Explorer personality. Explorers are described as motivated by a "systematic way to commercialize and scale," make decisions by "believing that every problem should be broken down into its constituent parts," take a "hands-on and directive" management approach, and have a leadership style that "attracts similar systems thinkers." Gaps common for Explorer personalities include loss of "interest after cracking the code," being "tough on team members," and "can become impatient with less sophisticated customers." Explorers' leadership styles are said to pair well with Crusaders and Captains but not so well with drivers or other Explorers.[3]

While *Built for Growth* provides details about each of the defined personality types, the authors also make the point that one type is not better than another. But the business leader's personality "is the animating force in building any new business" because their "combination of beliefs and preferences reflects his or her motivation, decision-making mode, management approach, and leadership style." In other words, the leader sets the tone, and his or her strengths, weaknesses, and preferences are the foundation that the rest of the team must build upon.

Kuenne and Danner agree that "these factors play out dramatically throughout the start-up and scaling of a new business."[4] I appreciate the authors' use of the word *dramatically* since uncertainty often does yield drama. Pressure only increases when a start-up takes on investment because an important stakeholder is added to the team. Investors are good at managing entrepreneur personalities, but team members may not have the experience to operate in an environment where power dynamics are evolving constantly. While positive financial outcomes are their biggest priority, investors also expect varying degrees of involvement, which can also add to drama.

So that we can talk more specifically about functioning in these environments, I'll start defining some terms since entrepreneurial ventures are formed with different goals and scorecards. Broadly defined, an entrepreneur is the person or team involved in starting a business using considerable initiative and while taking on risk. Blank and Dorf, who wrote *The Startup Owner's Manual*, define a start-up as "a temporary organization in search of a scalable, repeatable, profitable business model."[5]

Knowing that the team is temporary and, therefore, meant to evolve provides further context for early statements about the need to be comfortable with uncertainty. Some people enjoy that evolution while it drives others crazy.

Blank and Dorf go on to define four types of start-ups: "scalable start-ups, buyable start-ups, large company entrepreneurship, and social entrepreneurs."[5] Each of these types is expected to grow exponentially, but scorecards vary because funding and outcomes are different. For example, we expect scalable start-ups to replicate and buyable start-ups to be acquired.

Large Company Entrepreneurship will incubate a new disruptive product with possible outcomes such as a spin-off of a division. Finally, social Entrepreneurs will work to solve a social problem usually funded through grants or donors. Given the different types of funding and expected outcomes, their scorecards should reflect their ability to get closer to their respective desired outcome.

Another term to become familiar with is *lifestyle* business, which some entrepreneurs choose to support their financial needs and lifestyle. These businesses still require initiative, but the risk is lower because smaller investments are required. They are run with very few employees—often only one—and therefore have low overhead, making investments less necessary.

An example of a lifestyle business would be a graphic designer who simply requires computer graphic software tools to perform their work. They can do their work in a chosen location, likely from home, and with the ability to select the frequency and type of projects they prefer. They choose to prioritize projects and flexibility over the benefits that come with working full time for a company. Great entrepreneurial spirit and hustle are still required to generate business, but they trade off predictable income and paid vacation time. They are entrepreneurs on a smaller scale, and more women are choosing to start businesses in this space.

Entrepreneurs who choose the start-up path typically have an expectation to grow exponentially, whereas those choosing the lifestyle path are looking for something different. While a lifestyle business can grow to become Scalable or Buyable, these are usually not part of the entrepreneur's initial goal when establishing their lifestyle business. Goals are more oriented toward enabling flexibility and independence. These entrepreneurs work as much or as little as they choose, can turn down projects that don't fit their goals, and can work in the location they wish, all in support of the lifestyle they prefer.

THE CURRENT STATE OF WOMEN ENTREPRENEURS

Looking at the rise in women-owned businesses, current stats show that women own 42 percent of business in the US.[6] Rieva Lesonsky of SCORE says that 49 percent of businesses owned by women, whether they be lifestyle businesses or start-ups, are 0–5 years old suggesting a recent surge. Women-owned businesses fell into the categories of retail (22 percent), health, beauty, and fitness (16 percent), business services (13 percent), food and restaurants (12 percent), education and training (5 percent).

Lesonsky's also reports that women-owned firms employed 10.8 million workers in 2019 (up 28 percent from 2012). Women-own firms without employees now comprise 41 percent of all nonemployer firms in the country.[7] This data validates the trends I'm seeing when working with my own clients and in mentoring start-up companies. Many women are seeing entrepreneurship as a gateway to financial independence and flexibility, with the women-owned business category now representing an important component of the American economy.

MY ENTREPRENEURIAL EXPERIENCE

Taking myself as an example, I've pivoted between entrepreneurial companies for much of my career and thought it was illustrative to talk about what that was like. While these experiences were exhilarating, I wish I had known a few things as I was navigating these pivots, and I'll share those here.

Generally speaking, entrepreneurs don't come to this work without having prior business experience where they cultivate industry expertise and connections so they can identify where a new business opportunity may exist and then vet that with peers and then potential customers. My background began with spending ten years in the cable television industry doing consumer marketing, direct sales, and ad sales, and then I pivoted into technology communications joining my first entrepreneurial venture called Redgate.

Redgate Communications was a start-up marketing services company that focused on providing businesses with digital media solutions solving information distribution problems. Redgate had offices in Florida, New York, Massachusetts, and California. AOL bought Redgate in 1994, putting it in the buyable start-up category. Revelations from this experience include:

- **Sorting**—Leading up to the AOL acquisition, and in the absence of having any real details, some employees were spooked and left the company to avoid potential employment disruption. AOL was just becoming a known brand, so there was a lot of speculation about what this would mean for the Redgate team members. The confidentiality required while an acquisition is being negotiated is common to prevent any outside influence on a deal. This period can be unsettling to employees of the company being acquired. In my case, staying with Redgate required a leap of faith based on seeing a huge opportunity. As the AOL team evaluated us during the acquisition, it was clear that those Redgaters left standing were the ones comfortable functioning in the entrepreneurial environment AOL was trying to build.

- **Risk and savvy are required**—Once the acquisition was finalized, employees who wished to join AOL had to relocate to Northern Virginia. Those who did would have compensation that included some AOL stock options designed to balance out surprisingly low base salaries. Everyone who worked for AOL received some number of stock options no matter their job, making us all financially and emotionally invested. But having stock options didn't guarantee financial upside. The strike price of your stock options (driven by the market price when you joined) and whether you sold them at a high market price dictated your financial upside. Some people were more market savvy than others, and those who were stock literate did well. Internal lore said a maintenance employee had been with the company for over fifteen years and, therefore, had a very low strike price. Unfortunately, he never sold a single stock option and therefore, never realized any financial gain. Working under a compensation structure like this can be either exhilarating or devastating. I eventually learned to resist calculating the difference between opportunity and reality. It was better for my health.

AOL's Digital City was a start-up inside AOL, putting it in the Large Company Start-up category. AOL was working hard to foster innovation inside the company, and Digital City was one of those ventures. We created local online city guides that partnered with and competed with local print news and media companies. I launched Digital City Boston as market number two in 1996. Operations subsequently launched in many other major markets, including Philadelphia, New York, Chicago, San Francisco, and Minneapolis/St. Paul. Given that this was a start-up inside a big company, Digital City would fall into the Large Company Entrepreneurship category. Revelations from this experience include:

- **Challenges of recruiting an entrepreneurial team—** When building this type of company, speed to market was everything. As the mayor, I did not have a rule book for how to get the market up and running beyond high-level guidelines to find media partners and do deals, digitize their content, and sell digital ads around that content. The team we hired had to be comfortable functioning in an undefined environment. The best set of tips I've seen for finding entrepreneurial candidates came from *Harvard Business Review* and included a set of questions to consider about candidates. "Has the candidate made choices that clearly favor adventure and learning over convention and minimization of risk?" The article suggests making a note of things like "attending a less recognized college to pursue a passion; spending a year abroad in an unusual setting as a growth experience; opting to work for a highly innovative small company rather than a big brand-name company; vacation destinations that involve hardship but unusual experiences; living in a diverse and interesting part of a city rather than the usual professional enclave."[8] I quizzed candidates to get at these types of things back in the early 2000s and still continue to leverage them when recruiting candidates for myself or my clients.

- **Speed drives culture**—The need for speed contributed to a very competitive culture between the various Digital Cities. We couldn't be methodical; we were moving fast to secure additional funding for the venture. I quickly learned it was better to ask for forgiveness versus permission when trying something new. If the idea paid off and generated a lot of revenue, we would quickly roll it out in other cities. Had we been better organized, we would have prioritized each Digital City market differently. This would have prevented multiple markets from simultaneously trying to solve the same problems. For example, each market sold ads to car dealers, restaurants, and local corporations hiring new employees. Specializing may have avoided spreading each ad sales team too thin and helped us get smarter faster. This also ratcheted up the competition between Digital Cities versus creating an environment where everyone had a niche and shared expertise.

iBelong Networks created a software tool that helped membership organizations create branded, shareable multimedia content that could be easily published on multiple platforms (email, website, e-newsletter) while enabling member interactions. It provided functionality that didn't exist on corporate websites yet. Think of it as LinkedIn Business Pages meets Salesforce with your own company's branding.

I invested in this venture with a small number of other partners who each took an active role in the business. Had it survived, this company would have fallen into the buyable start-up category. We raised one round of VC funding, but we didn't move forward with the second round based on terms and, for me, some personal circumstances, which I write about in chapter 2: "Why Do We Pivot?" Revelations from this experience include:

- **Caution about dilution**—Preexisting investments are subject to dilution when new investors are added, which can change the power dynamic. The majority investor can make these types of decisions unilaterally. For the sake of example, let's assume three original investors contributing $10,000 (25 percent), $10,000 (25 percent), and $20,000 (50 percent) to a company representing a $40,000 total investment. If a new investor is added, contributing $15,000, then the ratios change to $10,000 (18 percent), $10,000 (18 percent), $20,000 (36 percent), and $15,000 (27 percent) with the new total investment of $55,000. (See the image titled "Dilution" below) When a company adds a new investor, it dilutes everyone's ownership percentage. The majority investor will always protect their majority share. In my example, the investor contributing $15,000 has guaranteed their greater influence over the two $10,000 investors.

Dilution

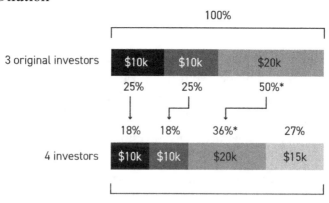

*majority share

- **Protect your interest**—Start-ups are messy and often function under loosely written or sometimes verbal agreements until the business is validated—a temporary organization, as mentioned earlier. Lean start-up methodology suggests that entrepreneurs should invest as little money as possible before validating their business—including the development of a minimum viable product (MVP). While business validation is the priority, my experience suggests that it's also important that team members communicate expectations. For example, do they want to play an active role, versus a leadership role versus an advisory role, should the business move forward. These may be difficult conversations to have at early stages, but they will prevent the drama that is inevitable if there are misunderstandings about involvement. Once investments start being made in the business, minority investors lose their ability to influence these and other decisions. There's a risk of becoming marginalized if expectations aren't understood. This may happen anyway, but it's important that everyone understands what they wish to get out of the venture.

- **The Impact of Investors**—Taking on investors is a huge step that propels your organization from a temporary state to a functioning company. Investment comes with expectations for hitting performance goals and lots of influence over decisions. Investors are likely to become majority shareholders in the business and, consequently, your new boss, which definitely impacts the personality of the business. I've seen investors insist on their choice of CEO, influence vendors and partners, remove a key member of the team due to a lack of confidence, and require rigorous reporting and processes. Start-ups should not underestimate the importance of building a successful working relationship with a new investor. Entrepreneurs are expected to be coachable, and those who are will benefit from the contacts and experience that investors bring. Those who are not will be miserable.

3WaysDigital is my marketing services firm that works with companies going through a growth transition. Examples of transitions can come from a spin-off, merger, new contract win, grant award, or new product launch—each requiring chief marketing officer (CMO) oversight and guidance to navigate their new circumstances. Since I don't have full-time employees, this would be classified as a lifestyle business. I either perform the work myself or provide oversight of a group of partners I trust. This balance was a logical and necessary pivot for me when our family was going through several transitions including our daughter's illness, our realization that our kids' childhood was zooming by, and changes in our day care situation. Revelations from this experience include:

- **Word-of-mouth marketing.** My clients have all come through someone who has worked with me before and who knows my skill set, making ongoing networking very important. You never know where your next client will come from so regular networking is very important.

- **Getting emotionally invested.** I get very entrenched with the C-suite executives, team and board members I work with. A typical project takes eighteen to twenty-four months, so I get to know people pretty well. Yet my task is to get them through their transition with a road map and a new internal marketing team so they can thrive on their own. It's often hard to walk away.

- **Interesting and intense work.** I've had the benefit of working with a diverse group of companies that are doing extremely interesting things. For me, I've found the excitement comes from learning about the new business and aligning marketing approaches to their circumstances and personalities. This tracks with my explorer personality. Because this work is so intense, I can usually only handle one or two large clients at a time.

- **Ramping up and down.** Unless timing works out perfectly between one project ending and another beginning, there's often a lag between client projects. This goes with the territory, and I've learned to embrace the down time between projects.

My entrepreneurial experience has helped me understand all that's involved in building something from the ground up and with special admiration for those who take this on. It's worth taking a deep dive into the experience of one woman who has built a billion-dollar company to show what it takes. While Anne Wojcicki is an incredible business leader having started 23andMe, we should acknowledge that she was incredibly well-resourced and connected. Let's not kid ourselves that people who build billion-dollar businesses bring vision, talent, drive, and some luck, but they always have incredible support networks that further enable them to succeed.

IN PRAISE OF WOJCICKI

Anne Wojcicki graduated from Yale with a bachelor's degree in biology. She evolved into the genomics space after working on Wall Street as an investment analyst overseeing healthcare investments. She became disillusioned by Wall Street's attitude toward health care and focused instead on biological research. She and her cofounders, Linda Avey and Paul Cusenza, began 23andMe in 2006 with the goal of providing a broader swath of people with access to their genetic data so they could make health and treatment decisions. Up until 2006, gathering genetic information was only available through specialists and at high prices.

While 23andMe wasn't the first company to market a DNA spit-test that sequences one's genomes, they chose to focus on screening for diseases while their competitor Ancestry.com focused on genetic identification. 23andMe started with a small team of five people and they faced several issues in the early stages. Given Wojcicki's background, she saw the value of people having access to their genome, but the market for people wanting to know whether they have markers for certain diseases had to be developed.

As it did develop, there were several controversies around the database itself. Test-takers could opt to share or crowd source their genetic information, making it possible for the database to be used for several purposes. Investors found this aspect of the business compelling. That database now has over twelve million records and is of interest to many different types of companies, such as pharmaceutical companies deciding what drugs to develop, law enforcement agencies trying to solve crimes, employers for deciding who to hire, and insurance companies trying to determine who to/not to insure. Some of those things have been deemed illegal, but upon founding the company it was open season.

23andMe determines who has access to the data and determines the price. A 2018 article in *The Atlantic* said, "The idea that customers are paying to have their DNA sequenced, only to have the drugs developed from that data sold back to them has sparked some backlash against the 23andMe and GlaxoSmithKline deal."[9] To provide an idea of the value of this data, the press release for the GlaxoSmithKline deal announced a $300 million equity investment in exchange for access to that database to "fuel drug target discovery."[10] According to Crunchbase, 23andMe has now received over twenty rounds of investor funding totaling $1.1 billion.[11]

Beyond the issue of who should have access to the data for large amounts of money is the issue of 23andMe being able to protect this information from hackers. Several pieces of legislation became necessary to protect constituents, which Wojcicki and her team cooperated with. The Verifying Accurate Leading-edge IVCT Development (VALID) Act called for government regulation and oversight of laboratory-developed tests. The Genetic Information Nondiscrimination Act was developed to protect Americans from discrimination based on their genetic information when seeking health insurance or employment.

After much consternation inside government agencies about which would be responsible for oversight of this new product category, the US Food and Drug Administration (FDA) was finally assigned the job. The FDA had to take a step back and define its responsibilities in protecting public safety in this uncharted environment before they moved forward with oversight. Everyone was charting new waters, and it was deep.

23andMe had to first develop the consumer market to create their database. Second, they had to develop the business-to-business market that was interested in tapping into their data. And third, they had to navigate the regulatory environment, which they seem to have done with appropriate transparency. I have great respect for Wojcicki's ability to navigate these waters with grace and integrity. Theirs was a long-term business strategy that required a lot of resources—and they had them.

Without detracting from Wojcicki's business prowess, and to my earlier point about resources and connections, Wojcicki happened to be married to Sergey Brin, cofounder of Google, when 23AndMe was founded. Google is still one of their investors. Her sister, Susan Wojcicki, was the CEO of YouTube, and her other sister, Janet, is a leading anthropologist and epidemiologist. 23andMe was financially well-resourced, plus they had great business and science minds very close to them. It's not unusual for companies with great ideas to run out of money before they can properly develop the business. Most entrepreneurs don't come to the table with a foundation like this.

For context and scale, it's worth mentioning that according to Rohit Arora of *Forbes*, the average revenue of a women-owned businesses in 2022 was $263,091. The average amount of funding a woman-owned business received was $55,898 versus $93,976 in funding for male-owned business.[12]

PIVOT LESSONS YOU CAN USE

1. **Not everyone is an entrepreneur.** Note the characteristics necessary to thrive as an entrepreneur to determine whether this is a track that's right for you.

2. **Start-ups evolve.** Understand that start-ups are temporary organizations that will take on many forms as they grow. Make sure you are comfortable with the evolution.

3. **Growing into entrepreneurship.** Prior business experience is often the training ground for identifying business opportunities and developing entrepreneurial thinking. Your relevant experience will be important to investors and to the team that you hire in your business.

4. **Identify your leadership style.** Understand which entrepreneur personality you bring to your venture and hire the characteristics you lack. The leader's personality type will drive the organizational culture.

5. **Design your opportunity.** Understand the drivers and scorecards for each different type of entrepreneurial venture so short-term and long-term goals can be established.

6. **Consider your assets.** When considering joining an entrepreneurial team, be clear about what you can give (time, money, connections) and what role you want to play. This isn't like going to work for a company that has a job description. The company gets formed around its key assets and then hires for what it doesn't have.

7. **Risk yields bigger rewards.** Companies that provide stock options are asking you to take the risk with them when you join an entrepreneurial team. That means strap in. Get educated on how to maximize your upside.

8. **Connections and resources matter.** Designing a billion-dollar company doesn't happen without connections and resources. If you don't already have them, allocate time to developing them.

Being an entrepreneur is hard work, but the financial upside can be significant if you have a higher risk tolerance. Serial entrepreneurs normally need a few months or even years in between ventures since the work is all-encompassing. For a dose of reality, Timothy Carter at *Entrepreneur* writes that 20 percent of small businesses fail in year one, 30 percent in year two, 50 percent by year five and 70 percent by year ten.[13] These numbers count businesses that no longer exist as a failure, which likely includes some that have "died by natural causes," especially in years five to ten. But it's safe to say that given the high failure rate, we should acknowledge the uniqueness of those who succeed.

6.

The Business of Mothering

———

As we walked around the quirky home with our realtor, she casually disclosed the house was soundproof. The owners never heard a thing coming from the basement apartment the night of the murder.

"Did you say murder?" Steve and I both asked.

The realtor was required to let prospective buyers know of the crime, which explained why this home was in our price range. In our search's early stages we suffered from sticker shock, given the greater Boston, Massachusetts, real estate market. Even though we were thriving in the Boston job market, we were questioning whether we could afford to raise our family there. We were not native to that area but had both relocated for jobs before meeting and getting married. We were house hunting in unfamiliar territory and were five hours from our closest family member. Despite it being in our price range, we did not make an offer on the murder house.

The consideration of proximity to family is one that will likely come up repeatedly as your family grows and changes. Steve and I seriously considered moving to the small city of Rochester, New York, not only closer to family but also in a much lower-priced real estate market. We spent several weekends driving from Boston to Rochester to interview for jobs with companies like IBM, Xerox, and Kodak. As we started getting closer to job offers, we started looking at homes and discovered the direct relationship between Rochester housing prices and the salaries of the area. We also realized that there was a limited number of companies we would be interested in working for. This experiment caused us to redouble our efforts to find a home back in the Boston area. No pivot necessary right now. We consciously traded off being closer to family in favor of better employment options and a more urban lifestyle. This was our early attempt at making one of those big-impact decisions that families find themselves making and wondering if they will regret later.

We did not regret this decision although there were times when life with children caused us to wonder what might have been if we'd made that move. I remember that thought coming up again when filling out an emergency contact form for our son's preschool. The form asks for the contact information of the three people the school should call if the child's parents are unreachable. Although we finally bought a home in a Boston suburb, the neighbors were new to us, and most were retired. Single friends in downtown Boston were not good emergency contact candidates. I fudged the form that year and set out to know more responsible, kid-friendly people in similar situations.

Managing the give-and-take involved in planning activities and building trust with new families requires time and effort. Data shows it is common for the mother to take on this type of role. According to a Pew Research study, "(54 percent) of parents in households where both the mother and the father work full time say that, in their family, the mother does more when it comes to managing the children's schedules and activities; 47 percent also say this is the case when it comes to taking care of the children when they're sick."[1] For those not living within close proximity to family, there will be moments when you may feel exposed, and close friends will fill those gaps once they are cultivated.

THE DAY CARE PHASE

Whether women work full time, part time, or are stay-at-home mothers, we all need day care at some point. Costs have risen dramatically since our children were born. If prospective parents look at these numbers too hard, they might get discouraged. In fact, many have since the number of children born to the average American family since 1960 has dropped from 3.6 children to 1.7 in 2018.[2] When comparing average US day care costs from the earliest data available from January of 1991 until July of 2023, prices have increased by 348 percent.[3]

When faced with childcare options, we no doubt consider whether our own extended family members should be part of the solution. According to Pew Research, "55 percent of US adults say they live within an hour's drive of at least some of their extended family members. One-in-five say they do not live near any extended family members."[4] While living close doesn't guarantee that these extended family members will be part of the day care solution, having a fallback nearby does provide a welcome emotional safety net in case of emergencies.

Any day care solution, whether it's from a family member or not, requires a nuanced relationship. Childcare providers are an extension of your family, and if all is going well, your children become very attached. Showing appreciation for the difference they are making in your children's lives should be ongoing, yet a power dynamic is at work, and I would argue that the power lies with the childcare provider.

As parents, we can make our expectations known, but it's impossible to anticipate every situation. The nature of the relationship is that you trust the provider's judgment to make good choices in the moment. They can make your child's day go smoothly or not—and that impacts your day too. If a child is sick or a care provider isn't at their best, walking out the door in the morning is *hard*. I found that compartmentalization was necessary to get any work done. But frequent check-ins were also needed for my benefit to know that things had smoothed out or that some intervention was necessary. For example, did our sick child need to see a doctor? Or did a situation the provider was dealing with (e.g., their own sick family member or a car problem) escalate or resolve?

Our family experimented with different types of day care solutions before finally deciding to hire an in-home nanny.

Despite the wonderful lifelong relationships we developed, this option is expensive, but we sacrificed other things to be able to afford it. It made so much more sense for us after we had multiple children with different schedules due to their nine-year age difference from oldest to youngest. The length of a typical school day being shorter than a workday started pushing us in this direction.

We also had an interest in making sure our kids could participate in after-school activities. The day care provider would meet the school bus, provide transportation to activities, or sometimes run an errand. Any pop-up events that deviated from the normal schedule would be discussed each morning as we did the fifteen-to-twenty-minute hand-off. This time at the beginning and end of each day was crucial to making sure that the kids had transition time, but also so adults could convey any important information to each other.

I mentioned my need to do check-ins during the day to deal with immediate issues, but planning check-ins with some frequency for longer conversations is also necessary. This is where you can discuss items such as potty-training progress, or any changes in the kids' routine. We could also brainstorm together how to influence the kids' behavior. This is a team effort. Steve was always present, and it helped if wine was served.

Having an in-home day care provider isn't where we started. Given our experience with various solutions, I tried capturing some pros and cons in the chart below to inform your own ideas about what you might be looking for.

Day Care Comparison Chart

PRO	Day Care Center	CON
• Almost always open with publicized schedule to plan around • Has programming/curriculum • Lots of socialization with other children		• Hard dropoff/pickup times • Exposure to illnesses • Price increases substantially when adding another child • The best ones will have a waiting list

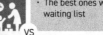

PRO	Day Care in Someone Else's Home	CON
• More 1:1 attention for your child • Homier environment • Usually the least expensive option		• Often syncs with school schedule because provider is usually caring for their school-aged children.(e.g., you may not have day care during the summer and school holidays) • Accommodations may be necessary since your child is joining someone's family/schedule (e.g., an undesirable pet, special house rules) • Price increases substantially with a second child • The best ones will have a waiting list

PRO	Day Care in Your Home with Au Pair	CON
• Can negotiate your preferred start/stop times • Can control the rhythm of the day • Potential foreign language skills		• You are giving up a part of your home, some amount of privacy, and taking responsibility for a teenager or young adult • Need to create house rules for weekend visitors, smoking, etc. Will need transportation for au pair • Au pair leaves after a year, and you have to build new relationships

PRO	Day Care in Your Home with Professional	CON
• Can negotiate your preferred start/stop times • Can control the rhythm of the day • Potential foreign language skills • Long-term relationships get built - they become part of your family • Provides continuity when school closes for holidays and summers		• Expensive if you need someone who drives • May need to provide a car that's safe and reliable for your kids • You are managing an employee. This is tricky business because you are doing it in front of your kids. Requires a level of finesse and devotion you shouldn't underestimate. • They eventually leave, and it can be traumatic for everyone involved

ONGOING CONSIDERATIONS

The normal logistics that come as children get older caused us to reevaluate our day care situation and increase our costs to accommodate the added complexity of three children and different activities. We continuously revisited this expense, but we saw a downside to having one of us completely leave the workforce. We kept coming back to these four considerations:

1. This is a temporary stage, and we are investing in our long-term careers. Note that "43 percent of highly skilled women leave the workforce after becoming mothers."[5] The percentage of men who leave the workforce "is still quite small but has increased at a remarkable rate, from 1.6 percent to 3.4 percent of all stay-at-home parents in the first decade of the new millennium."[6] The reason that either men or women leave is almost always to care for someone and report finding it to be a transformational experience.[7]

2. We both knew from our separate stints at home full-time with our kids—me during three maternity leaves and my husband during his year as a stay-at-home dad—that we both enjoyed *quality* time with the kids. But we craved the external stimulation and independence that came from working. Families need to determine what works best for them and be prepared to constantly reevaluate the decision as life happens.

3. We were willing to sacrifice elsewhere to make this choice—so no extravagant family vacations, fancy cars, or club memberships.

4. We would reconsider this choice if there was ever an issue with one of the kids, if our financial situation changed, or if something happened with our nanny.

GRATEFUL

We worked with three different women who were a significant part of our children's lives. Julieanne was with us for a few years after our oldest was born. She brought her son to our home each day, and the boys were each other's favorite companions. While I was on a three-month maternity leave with our second child, we couldn't justify paying someone to come to our home while I was not working. Not surprisingly, hitting pause on that relationship required her to find another family. We were very sad but totally understood.

Losing Julieanne led us to work with Raquel once I went back to work. She was with us for several years until we moved to Northern Virginia, triggering a tearful goodbye. Our two older children had a role in Raquel's wedding, and she later named her daughter after one of ours.

Agi was with us for nine years following our relocation to Northern Virginia, and she became very integrated into our family. Our kids also played a role in Agi's wedding, and she remains especially important to our youngest daughter, who was born during the time Agi was with our family.

These three women played important roles in all our lives and were complete strangers prior to us hiring them. I can't speak highly enough of the richness they added to our family. We consider them friends for life.

LONG-TERM VIEW

In talking with Dr. Marian McKee about the idea of investing long term in your career, she shared a story about how she also struggled with balancing. Marian juggled getting her PhD in Microbiology and Immunology with raising her family. Marian is currently the vice president of biosafety at Eurofins BioPharma Product Testing in Lancaster, Pennsylvania. During the time she was working on her PhD, she got married, had two children, and suffered the passing of her father.

Her husband traveled extensively for work, and Marian says, "I was paying for day care for two children and going to grad school. We were clearing one hundred dollars off my research stipend. I looked at my husband and said why should I bother? Maybe I shouldn't do this. My husband's response was to say that we are investing in the future." With his encouragement and that of her professors, Marian finished her PhD and is now reaping the rewards of her hard work.

In other evidence of the importance of a long-term view, Marian shared how the achievement of her PhD at age thirty-three delayed her ability to tap into higher corporate salaries. For several years, her salary was lower than that of some of her less-credentialed colleagues who had started working immediately following their undergrad experience. Eventually, the combination of Marian's work experience and her advanced degree allowed her to achieve the salary levels she was entitled to, but this took some time. A long-term view was something that Marian was familiar with, and she summoned this experience again when reconciling her early salary levels.

During those moments when we ask ourselves whether we should keep working outside the home, it's important to consider the impact of leaving the workforce. Heather Gillin of *Texas A&M Today* writes, "Women who opt to reenter the workplace after an interruption, like leaving work to raise a child, often face a downward career trajectory." Those who return to work often suffer consequences such as being "expected to choose lower-level positions, less prestigious occupations or care-oriented professions with lower pay to balance work and family responsibilities." Gillin's piece also says that "factors like length of parental leave, full-time or part-time status before having children, number of children, age at first birth and education level all influence wage penalties."[8]

Neither Marian nor I would have been successful juggling childcare and career without our partners' help and support. This only works if you have a partner who actively participates. As already mentioned, my husband Steve stayed home with the kids when we first moved to Northern Virginia. He has funny stories about being the only dad on the playground and getting skeptical looks from mothers who didn't know how to react to him. Interestingly, a former aviation software colleague recognized Steve when he was performing kids' reading volunteer duties at our local library. When her company had an opening, she reached out to Steve and eventually hired him.

WATCH OUT FOR POACHERS

When you have a great day care person shepherding your kids to after-school activities, other mothers take notice. Some began asking our nanny to babysit for them on evenings and weekends. I took this as a testament to how competent and wonderful she was, but it started to become invasive. On one occasion, a mother asked our nanny if she could drop her kids off at our house while the mother attended an event. In another case, a mother asked our nanny to arrange to leave our house early to allow her to make an evening commitment.

Since these were acquaintances and neighbors, our nanny was baffled about how to respond without offending anyone. Luckily, our good working relationship was strong enough that we could discuss these things as we transitioned in or out. It was awkward because I certainly didn't want to prevent our nanny from making extra money outside of our agreed-upon business hours, but I was stunned that others were not respecting boundaries. Ultimately, I contacted these mothers directly to take our day care provider out of the middle.

As a full-time working mother, the relationship you have with your day care provider is sacred. You may find that you must protect it from those who don't understand how much of a lifeline this really is.

PIVOT LESSONS YOU CAN USE

1. **Control freaks relent.** It's hard to add a new day care person to your family in whatever form that takes. You will never really feel in control—because you aren't. You are trusting someone else with your kids' well-being. Develop an appreciation for the great benefits that come with someone introducing safe and appropriate alternatives.

2. **Temporary state.** Be prepared to constantly reevaluate your day care decision. Both internal and external forces shift simultaneously so adjustments will be required.

3. **R-E-S-P-E-C-T.** Treat your day care provider with the respect and inclusion they deserve. While they work with you, they are an extension of your family. How you treat them affects how they treat your kids.

4. **Transition time.** Whatever type of day care option you choose, transition time will be necessary from parent to provider to parent each day. Crucial information will need to be conveyed, and the child will need time to settle into this next place physically and emotionally.

5. **Open communication.** The nuanced relationship between you and your day care provider requires open communication to last. Consider monthly meetings over a glass of wine to have those "how is it going" conversations.

6. **Consider the long term**. At times you will ask whether staying in the workforce is worth it. Trust your gut but consider the setbacks that are likely once you reenter the workforce.

7. **Keep the good ones**. Once you find a great day care person, do what you can to hang on to them. They will likely be part of your family long beyond the time you are paying them to care for your children.

8. **Poaching is real**. Others will want access to a great day care person. Consider what boundaries you want to put in place and prepare to make those clear.

With three children now grown, I'm far enough away from the days of juggling day care and a full-time job that requires being in the office every day. I also acknowledge that things have shifted in this post-COVID-19 environment. The need for parents to be in the office for eight to ten solid hours changes the day care game somewhat.

Parents are no longer stigmatized if they must stay home with a sick child. Cameras can stay off during Zoom meetings when kids are home sick. The amount of research on the challenges that working women face encouraged the creation of safe spaces for discussion and employee assistance programs. Companies are putting better support structures in place so women can turn to each other and to their supervisors when things get challenging.

Women will always walk a delicate line when trying to achieve big things professionally, especially while managing families. I imagine some of the same principles discussed in this chapter apply to elder care. These relationships are much more delicate and tenuous than many realize. This phase does pass, but not without constant adjustments. The current shifts we are seeing in corporate support are a welcome acknowledgment of the realities of parenting/caring and the importance of women in the workplace.

7.

Unexpected Ways of Impacting Personal Brand

How do people find those great mentors? Envy is a counterproductive emotion, but I'll admit to feeling it when people say, "my mentor blah, blah, blah..." In my experience, those who claim to have mentors are often post-docs referring to their doctoral advisers. Those very formal relationships tend to last for years, even after the candidate has written their dissertation, due to common interests and the intensity of the work they did together. These mentoring relationships bring accountability on both sides and are full of mutual respect.

For the rest of us, these types of relationships seem very rare. According to *Forbes*, only 37 percent of professionals say they have a mentor. Of those who do have mentors, 61 percent say the relationship started naturally, and 14 percent say the mentorship started when they asked for mentorship.[1]

Given that there seems to be an organic component to mentoring, perhaps an openness to the idea is a first step.

The changes in modern-day employment relationships are a big factor in why formal mentoring programs have become less common. Since employees aren't staying with companies nearly as long, nor are they even working for one company at a time, loyalties on either side of the mentoring equation have shifted, except in Fortune 500 companies, where 70 percent report formal mentoring programs are still part of their culture.[2] An argument can be made that this shift in loyalties is precisely the reason why mentoring is still important for companies, but let's look at why mentoring is important to individuals.

Several studies point to positive benefits for people on both sides of a mentoring relationship. In a *Harvard Business Review* study of new CEOs working in large corporations who had access to seasoned counsel and feedback, 84 percent said they became proficient in their roles faster. It went on to say that 71 percent were certain the company's performance improved as a result of the mentoring.[3] A Wharton Business School study found that 25 percent of employees who took part in a company's mentoring program had a salary grade change, compared with 5 percent of employees who did not participate in the program. It also found that 28 percent of mentors had a salary grade change as opposed to just 5 percent of those who were not mentors.[4]

Other, less quantifiable benefits cited by those who mentor include things like the experience provides them with an ability to gain additional company insight that they may not have access to otherwise. Also mentioned was a general satisfaction as well as an ability to see how others perceive you as a leader, an ability to create larger networks that you may need later, and finally, to gain more experience resolving issues.[5] For those on the mentee side of the equation, notable benefits include help defining career goals, the improved accountability that comes from successfully completing tasks meaningful to them personally, and the ability to have confidential conversations.[6] One can see how each of these things can also contribute to a company's bottom line.

Suppose your company doesn't have programs that help them identify and develop the superstars in their organization. In that case, you may need to be more aggressive about seeking out your own mentor relationship. Your company would likely perceive this as a demonstration of loyalty and an interest in moving up in the organization. It also demonstrates self-advocacy. For anyone who has played or coached organized sports, the coachable athlete is also memorable simply because they are eager participants. This idea holds true in the workplace as well. But how do we start?

PROFESSIONAL DEVELOPMENT ORGANIZATIONS

Another option if you don't have access to formal mentoring is participating in a professional development organization. This also signals to your company an interest in moving up, especially when you bring learnings and experiences back to your employer. These organizations are often gateways to finding mentors or being a mentor since many feature those types of programs as member benefits. Of course, these organizations also provide access to new ideas and people. We've all heard that what you put into these experiences correlates to what you will get out of them, but first, let's understand the different types so you know where to look.

Not all organizations serve the same purpose. According to Claude Balthazard, writing for LinkedIn, there are four types of professional organizations:[7]

1. **Member benefit professional associations**—provide access to resources and support through advocacy, networking, organized events, products, and services

2. **Designation granting organizations**—similar to associations but also offer professional learning and designations

3. **Certifying bodies**—focus is on designations as a product, sometimes prior education or professional experience is a requirement to participate

4. **Professional regulatory bodies**—promote the public interest, not the professionals under regulation

Depending on the industry you are in, certifications may be required to progress to higher levels. Your company would communicate the need for such designations, and therefore, participation in type two or three might even be sponsored by your company. Since type four focuses on the public interest, credentials may be necessary to participate at a high level. Examples of these regulatory bodies include the Department of Health or Department of Environmental Protection. Most would have volunteer opportunities available for the benefit of the public good.

Benefits of type one require a thoughtful participation strategy to reap the greatest dividend. In Member Benefit Professional Associations, assuming volunteer roles provides relevant experience and access to leaders of thought and of programs. Participation would also provide access to the formal mentoring programs that many of these organizations have. My experience with Women in Technology (WIT) is an example of leveraging volunteer opportunities for personal growth, which I discuss in the next section.

MENTORING AND VOLUNTEERING

I recently read Geena Davis's new book *Dying of Politeness*, where she talks a lot about the soul searching she did to find her own identity. Geena speaks of not really having a female role model until she met Susan Sarandon on the set of *Thelma and Louise*. "She was someone who says what she thinks and knows what she thinks and is very comfortable in her body and moving through the world. I realized that I had never spent extended time with any woman who was like that, who simply said what she thought. It was so valuable to be able to spend that time with her because I learned what I wanted to be like."[8] Davis goes on to say that "I kicked ass on-screen way before I did so in real life."[9]

The title of Geena's book comes from her learned behavior to be polite even when people crossed appropriate boundaries and made her feel uncomfortable or, in some cases, inept. In her book, she provides countless examples of how she was misunderstood by many of those she worked with in films. Her politeness made others feel less awkward and prevented Geena from being perceived as difficult. This sense of always being pleasant and accommodating was also ingrained in me. Making another person, especially a man, feel accountable was something I had little experience with.

Davis's acknowledgment of the impact of women mentors like Susan Sarandon reminded me of some of my own experiences working with girls exploring STEM, where we focused on making sure girls could see what they could be. STEM stands for science, technology, engineering, and math. Recently, an *A* has been added for arts, so you may also see references to STEAM.

This movement over the last twenty years works to make sure girls pursue STEAM education since there is destined to be a shortage of experience if they don't. According to a study done by *Science*, "Common stereotypes associate high-level intellectual ability (brilliance, genius, etc.) with men more than women. These stereotypes discourage women's pursuit of many prestigious careers. Stereotypes are endorsed by, and influence the interests of, children as young as six."[10] *Science* also suggests that we must catch girls early to impact their interests. Davis's book agrees that we must catch girls early to impact their self-esteem. I found both things to be alarming given my volunteer experience with WIT where I watched girls second-guess themselves.

I've now been a member of WIT for fifteen years, having participated at all levels of the organization, including as president. WIT is a type one, Member Benefit Professional Association, providing broad support for women at multiple points in their careers. WIT programs include mentoring, training, networking, and recognition. As with many of these types of organizations, WIT's committee structure is the way work gets done. Each major initiative is run by a committee of volunteer leaders and members who are passionate. They become the engines that propel important initiatives forward.

When I first started with WIT, I chose to participate in the Girls in Technology (GIT) committee. At the time, my oldest daughter was choosing her high school electives, and we had begun noticing her being apprehensive about taking an AP chemistry class. She loved science, so this seemed out of character. This shift was something we had not experienced with our son, and I wanted to learn more about how other girls were handling these decisions.

The GIT program provided high school and middle school-aged girls with monthly STEAM-focused mentoring sessions that helped them explore their interests in STEAM, build confidence, develop strategies for success, and meet women who were currently working at the highest levels of STEAM careers. The see-what-you-can-be—and improve upon—approach was playing out through this program.

As an example of how these organizations provide opportunities to learn and grow, I want to paint a picture of the work we did as a committee. The committee sourced professional women speakers and mentors, screened applications from the prospective protégés, matched protégés with relevant mentors, and organized monthly events that brought all the constituents together. Each session had a topic of discussion and then designated time for the one-on-one mentoring to happen.

We worked to find sponsors who would donate meeting space and ran a scholarship program so girls could fund further education. Pulling off these programs each month felt like a triumph. As volunteers, we had a common mission to make each of these sessions meaningful for the girls, and we problem-solved logistics and tweaked for improvements as we went. The protégés were insightful and eager. They were also very busy.

As you might expect, these girls tended to be overachievers who were looking for a leg up as they approached college applications. Seeing other women who had applied STEAM educations to real-life careers was an important frame of reference for the girls. Volunteers also benefited from hearing the speakers' stories of their challenges and tricks for making it through engineering programs and how they used those skills in their day-to-day work.

Out of this experience, I gained several insights that I wasn't finding in my daily work or as a parent for that matter. I joined the leadership of Girls in Technology and later the Board of Women in Technology, and the experience taught me several things including:

- Passion for a worthy topic brings very different people together to do great things.

- Girls need different support and guidance when making decisions about education and career because impostor syndrome starts at a young age.

- A few key individuals drive programs like this and without them, the programs wither.

- Breaking tasks down into opportunities of different sizes is the best way to leverage volunteers. Not everyone is able to play big roles, but some would do smaller ones if available.

- Volunteers can accomplish a lot when programs are small. Once you hit a certain size, you can only create something of quality when you pay someone to focus on it.

- Learning to tap into volunteers' strengths and passions provides great rewards for everyone involved.

- Corporate funding comes with agendas that sometimes conflict with a philanthropic mission, but you can't have one without the other.

I would put my WIT/GIT experience on the list of the ten most memorable of my career. The entrepreneur in me was constantly looking for ways to capture the best parts of GIT and scale it. We wanted to bring it into the more needy communities of Washington DC, Maryland, and Virginia and experimented with doing so. It became obvious that higher-level funding would be required, and applying for grants was the best way to grow. Corporate sponsors required yearly renewals that were subject to shifts in corporate priorities. Applying for the type of grants we needed to ensure long-term success required shifting the organization's legal status. The WIT board ultimately voted to make this shift and has now begun the journey toward seeking grants to expand upon programs like GIT. This was ten years in the making.

We do not take on these types of endeavors unless we are passionate about the outcome. Women negotiating a pivot will leverage all parts of their network to support them during these times. Connections from volunteer circles can be particularly valuable because they are not industry based and, therefore, extend network connections into new industries.

Finding programs that align development opportunities with things going on in your personal life makes them the best fit. An avid runner I know started a Girls on the Run chapter at our elementary school and later joined their board. This also became a big part of her personal brand. You'll find opportunities to become involved in all sorts of places, such as your kid's schools, places of worship, and community organizations. While these may not be official professional development organizations, they offer similar experiences and skill development.

Consider these questions when making decisions about organizations you wish to become involved with to determine whether they are worth your valuable time:

1. Does the organization and topic align with my values?

2. Does it provide an opportunity to flex different muscles than I'm using in my day-to-day work?

3. Does it provide a path toward recognition that raises my profile in a positive way?

4. Will I meet people who expand my network in new ways?

Time/Value Considerations

Endeavors that support personal values, expand skills and networks, and layer in recognition opportunities are excellent ways to facilitate your personal brand.

PERSONAL BRANDING

I consider mentoring and supporting women and girls in technology careers to be a big part of my value system and, therefore, my personal brand. Many books have been written about thinking of yourself as a brand and then leveraging product brand strategies to strengthen and promote the brand. At its core, the concept of personal branding starts with identifying the values that distinguish you from others and which provide insight into how you approach your job and your life. These aren't skills but beliefs that guide you and add color to who you are and what you bring to your roles. Werner Geyser of *Influencer Marketing Hub* suggests that you should be able to condense your brand down to a single statement that is "strong, descriptive, short, and catchy all at the same time."[11] For example, the current rendition of my personal brand statement is Fractional CMO for Growth Transitions: A Mother, Speaker, Mentor and Author Who Leads." One's personal brand statement may evolve over time, but capturing the essence of who you are is the first step.

The best examples of great personal brand strategies involve showing versus telling through repetition. This improves understanding and aids retention. Creating a repository for the thoughts and ideas related to your personal brand is a challenge people solve in different ways. Creating posts of work samples, relevant events you participate in, or volunteer activities you engage with on social media are all great ways to reflect your personal brand.

Different industries also provide publications and conferences where people can share thoughts and ideas through speaking and writing. Some advanced personal brand marketers purchase their domain name and set up a simple website. For example, trishbarber.com currently reflects the work I'm doing on this book, speaking topics and engagements, other projects that I'm attached to, and my blog with ongoing thoughts and ideas. Such a repository that you can easily refer people to reinforces your brand.

Geena Davis's brand evolution can be seen in her memoir, and I've watched it with interest throughout her career. Knowing her first as a model turned actress in one of my favorite movies, *A League of Their Own*, placed her in a performer category for me. I didn't think of her as a visionary until I learned more about her work founding the Geena Davis Institute on Gender in Media.

This foundation was the result of Geena watching children's films with her young daughter and suddenly realizing how few female characters there were in those films. She dug into this further and began quantifying the ratio of male to female characters. Geena says, "So what message are we sending girls and boys at a vulnerable age if female characters are one-dimensional, sidelined, hyper-sexualized, or simply not there at all? We are saying that women and girls don't take up half the space in the world. We're teaching them to have unconscious gender bias from the beginning."[12]

Her foundation website says it is "important that children see diverse, intersectional representations of characters in media to reflect the population of the world—which is half female and very diverse—and avoid unwittingly instilling unconscious bias in them."¹³ Her personal brand has evolved, and she talks about this evolution in her book. She didn't see herself as a visionary either. That happened over time. I now think of her as a credible visionary supporting women's issues through a data driven and collaborative approach.

Geena is still being polite. Her approach has been to work collaboratively with the entertainment industry by presenting them with the data she had collected to bring about the change she was envisioning. Her approach was exactly right for the time, and because she was a credible insider of Hollywood, it worked beautifully. Two of their primary goals were achieved in 2019 and 2020, which saw gender parity for female lead characters in the top 100 largest grossing family films and the top Nielsen rated children's television programs.

I stumbled into mentoring and volunteering through my experience with WIT/GIT and it tied together things I was learning with things I was seeing firsthand in the tech industry and inside my own family. Mentoring and volunteering now defines me, and my business is built on this concept. As a fractional CMO, I'm mentoring a group of people as they work through a big transition and then leaving them equipped to take on the task without me. I've now become very focused on increasing women's participation on corporate boards because companies haven't yet realized the value that women can bring to the leadership of their companies. But that's another book.

PIVOT LESSONS YOU CAN USE

1. **Mentoring impacts everyone involved.** You will find personal growth and benefits from mentoring others. Companies don't always have organized mentoring efforts, so you may have to find opportunities elsewhere or be a catalyst for change.

2. **Flatter someone.** Looking for a mentor? There's nothing wrong with asking someone to be a sounding board/guide/adviser. They will likely be flattered.

3. **ProDev is a gateway.** Find a professional development organization that provides access to people and ideas you care about. It will be invaluable during your next pivot.

4. **Regulate.** Look for opportunities that fit within the amount of available time you have to spend. The company you are working for will appreciate you taking the initiative and will also reap the benefits of what you learn.

5. **Market your brand.** Start thinking of yourself as a brand and invest time in building that brand. This will seem daunting at first because it requires soul searching. Start with social media posts at some manageable frequency, and eventually it starts to become part of your business day/week/month.

As Geena Davis says, "But if what I've written here has led you to believe that I have become a certifiable, full-time badass, don't be fooled."[14] My perception of her has completely changed over time, and I now think of her as a relatable and brilliant spokesperson on women's issues in Hollywood—a badass. She would argue she is still dying of politeness, but that's also a wonderful part of her personal brand.

8.

Networking

Running for office was a full family business. When Dad ran for a seat on the Board of Supervisors for our town in New York, my mother, my two siblings, and I joined him on the campaign trail at local pancake breakfasts, clam bakes, and spaghetti dinners. I remember him giving us lessons in firm handshakes, making direct eye contact, and delivering a respectful greeting to those we were trying to impress. I didn't know it then, but we were learning the basics of networking, and I hated every minute of it!

I have terrible memories of being pushed to say something witty when I couldn't think of anything but could respond to questions, "My favorite subject is science, and yes, I like pancakes." I hated the pressure of saying something that might negatively influence his success and felt like the stakes were high. I didn't like having to be charming on demand. I was not good at pretending to care about things I really didn't care about. I managed to push through the campaign events by adopting a "fake it till you make it" strategy—long before I ever heard the now-common phrase.

That introversion showed up later in life when I took the Meyers Briggs Personality Test as part of a college class and then again later in a job. I tested slightly different each time, teetering between an *I* (introvert) and an *E* (extrovert). My dad, being the gregarious guy who got his energy from being around others, saw big deficiencies in people who weren't comfortable commanding a room. I think my unconscious bias had something to do with why I tested differently. Being introverted was not considered an admirable quality in my family.

Fast forward to being on the campaign trail of my own life. Any networking event I attend can trigger old emotions of Dad's expectations. Heart pounding, face flushing, and tongue-tying can occur until I get a read on a person. I rely on the firm handshake and good eye contact skills my dad taught me way back when, but I have added in the use of open-ended questions that get the other person talking.

I'm great at listening, and I'm genuinely interested in what other people think. I'm good at connecting topics of conversation, making observations, and keeping people engaged. I always learn something new that I go back and research and often think about these topics for several days. Feeding my Explorer personality now makes networking fun. What makes it not fun is encountering someone who has an agenda and becomes dismissive if it is not being served.

NETWORKING BEHAVIORS

I've networked in lots of different circles and have noticed a big difference in the predominantly male groups. According to an article in *Forbes,* "Men and women have very different brains, and they do network differently. The male brain is more compartmentalized. They get straight to the point. They know the goal. They tend to decide right away, with little to no small talk, whether they will work with you or not."[1] This has been true in my personal experience. I've been in situations where someone has clearly determined I didn't have anything of value to offer, and I've noticed them looking over my head for someone else to talk to, no doubt someone more likely to serve their purpose. I'm happy to let these people move on.

Conversely, women tend to take a less *transactional* approach to networking. Caroline Castrillon of *Forbes* agrees and writes, "Women generally hesitate to ask for what they want out of a networking interaction. Instead, they think about what they can do for the other person first."[2] This again reflects my own experience when networking in predominantly female environments. Generally, there is a lot more inclusion.

There is always talk of professional topics but often talk of family and relationships. It's not unusual to hear a woman say, "I almost didn't make it to this event because my child got sick at school today…" after which the conversation might veer into something more personal. I've been that woman, so I'm quick to empathize when I see others in this state. "That's so unsettling. Clearly, you worked hard to be here, so what are you hoping to get out of this event?"

The behaviors mentioned here and attributed to males or females are meant as guidelines since circumstances are always unique. But knowing about these general differences will help you be less thrown off when you see them. Regardless of the environment, I've always found that going into a networking situation with a curious mindset helps me move through the experience more seamlessly.

In fact, my approach to networking is like that of my approach to shopping. I know that parking may be a hassle, there may be lines at checkout, and my size may be out of stock, but I go with a general curiosity and excitement about what I'm going to experience. I go alone so I can pick the stores I want to visit. I allocate enough time to make the experience relaxing. I take note of what I learn and bring those back into my own life. I also treat myself to a fun coffee to make the experience even better.

Applying the same logic to networking, I go alone so I'm free to make a conversation last as long as I'd like, and I can also leave if it's not clicking. I make sure to allocate enough time for transportation and locating the event so I'm calm and present when I arrive. I take note of the things I hear and the people I meet. Of course, I treat myself to a fun beverage that makes it all worthwhile. As someone who practices yoga, setting an intention for each practice is common. Intentions can be about learning, openness, friendliness, bravery, or anything that makes sense to you at that moment. Try setting an intention for your next networking event.

A *Harvard Business Review* article highlights that for people in power, networking is often easier because they are more confident about having something to give to others. The article encourages us to think beyond power about what else we can give in these situations. Many people "focus on tangible, task-related things such as money, social connections, technical support, and information, while ignoring less obvious assets such as gratitude, recognition, and enhanced reputation."[3]

A heartfelt comment to someone for sharing a unique perspective can make someone's day. I've also given compliments to the event organizers when I thought something was particularly well done. Then, I use the idea in future events I am involved with. At one event, a woman provided me with a lead on a vacation home owned by her friend after we had a conversation about my interest in that destination. I followed up with her afterward, and we rented the spectacular home a year later. The person I met was thrilled and so was the homeowner. Steve and I loved the vacation we planned around that wonderful home. It was win-win-win.

Your presence at these events is meaningful, and your perspective is unique. Capitalize on opportunities to give and take. Don't underestimate the power of your contributions.

BEING PREPARED

Because of the different styles of networking that you will encounter, a bit of preparation is in order. Borrowing from the more male characteristics of networking, think through desired takeaways. You've chosen to be at this event for a reason, whether it's impressing a prospective investor or simply learning more about the topic at hand.

Remember to give thought to what you want people in this ecosystem to know about you.

During a recent Zoom event, I was networking with a new peer group of well-connected academics and entrepreneurs. Being the newbie, I was asked to introduce myself. In addition to my professional background, I included a bit about the issues I cared about: mentoring women in tech and girls in STEAM as well as increasing the participation of women on corporate boards.

Immediately following the meeting, I got an email from one of these new colleagues who had taken note of the STEAM part of my introduction. He was inviting me to lunch with a STEAM-related organization he was interested in me being involved with. Had I not mentioned my extracurricular areas of focus, it could have been a missed opportunity for both of us.

Once, I heard someone introduce themselves by saying their name and "I help good companies become great companies." This statement had a lot of holes in it and raised some eyebrows with intrigue. It didn't really tell listeners what role she played or what kind of companies she helped. It did get people to ask more questions, and the discussion moved along nicely, but I wouldn't leave so much to chance.

Being vague or only providing the name of your company and title isn't descriptive enough, especially if you are networking with people from outside the same industry. Be descriptive about your role by saying things like "I think about… and I'm here to learn more about…" or "I focus on making sure that… so I'm hoping to meet people who…" or "my biggest priority is… and I'm eager to find…" These types of introductions make it clear to the listener what you do and how they can help you.

At times a group you are networking with is just not clicking. The event may not be well-attended, a key speaker didn't show, or it's the wrong audience. For example, I've been to events that brought together entrepreneurs with venture capitalists to keep tabs on the local entrepreneur community and potentially find a client. These events were designed to facilitate people meeting for the purpose of getting or giving funding. Since I was not on either side of that transaction, those I spoke to moved on quickly to find those who were. Early-stage companies didn't have an immediate need because they needed money to support marketing services. VCs looking to build their portfolios were not the right people, but some took my card for another time. These events were good for broadening my understanding of the ecosystem but did not provide short-term business opportunities.

CONNECTING WITH PEOPLE

So you've attended an event, learned something new, or met someone interesting, but without taking some action, these meaningful experiences aren't being maximized. But some follow-ups are more obvious than others. In the case of the vacation house I mentioned, I was motivated to know how much this vacation home cost. Here's the follow-up I sent: "We met last week at the XYZ event, and you mentioned your friend's vacation home. Would you mind sharing a link to the house details? I'm interested in learning more about the property." Pretty straightforward. But let's talk about ways you can determine what the best follow-up might be.

Your answers to the questions captured in the image below titled 'Evaluation Questions' help you gauge the follow-up actions you may wish to take. Actions should consider how much further you wish to take the relationship, how much value you got from the encounter, and how relevant it is either personally or professionally. At a minimum, for people you have meaningful conversations with, I recommend connecting on LinkedIn. Make sure to include a contextual note saying something like, "We met last night at the XYZ event, and I enjoyed our conversation about ABC. I'd like to add you to my professional network since ABC is a top priority for me."

For those people you didn't speak to directly but may have heard on a panel, a contextual note saying, "I heard you talk last night about ABC at the XYZ event, and I thought your point about DEF was right on target." Notice I said "last night." People are used to having LinkedIn follow-ups after attending a networking event, so they will be more open to accepting your request to connect in that context. If you wait too long to connect, the window will have passed, and your success rate will go down.

As a next level follow-up for important connections, you may wish to find further low-risk reasons to continue discussions. Perhaps invite them to attend another contextual event after you've made your initial connection on LinkedIn. The next communication can be something like, "I wanted to make sure you know about the upcoming UVW event. I plan to attend and would love to talk further with you there." Or "I have some additional thoughts about ABC, and I'd be happy to share my experience with you since I know this topic is important to you. Can we schedule a coffee?"

Evaluation Questions

IDEAS	PEOPLE
Does this idea align with your personal brand or a current objective?	Does this person work at a company you want to know more about?
Is this something you wish to research further?	Would others in your network benefit from an introduction to this person?
Is this something you need additional perspectives on?	Is this person attached to an idea you have interest in?
Are there champions of this idea that you should know?	Do you have thoughts/ideas/ support to offer this person?

If appropriate, you may wish to make the next step a specific *ask*. In this case, your approach should take as much work away from the other person as possible. For example, "I'm interested in a job at your company in the CBS department, and I see that Bob Jones is the head of that group. I wonder if you think he is the right person to ask for an informational interview to determine his group's hiring needs. Some of the current job openings I found online seem like a match for my skills, and I'd greatly appreciate the introduction." In this case, you've done some of the research and are making a specific request. People are too busy to try and read into what you might be looking for, so I've always found it best to be direct. But do what feels authentic to you.

CONNECTING WITH IDEAS

When following up on ideas that are of interest, it's helpful to start by letting your professional network know you are thinking about a specific topic or task. Again, I'd recommend LinkedIn, and I might start like this: "After attending XYZ event, I've been thinking more about how... and my observation is... and I'm wondering if others agree?" Include a link to the details page for the event you attended and its hashtag.

Event organizers track #hashtags to monitor social posts after events, and there is a good chance they will amplify yours. It would be interesting to see who engages on the topic. Request to connect with any folks new to you who engage with your post. Something like, "Thanks for your interest in my post. I'd love to connect." At a minimum, your network is on notice that this is a topic of interest for you, and they will begin to associate you with it.

If this is something you might be looking for further resources around, consider approaching your professional network with a more specific request. "After attending the XYZ event, I'm looking for additional resources on the ABC topic. Does anyone have data on… or know people who…" Adding links and hashtags provides more context for those interested.

Make sure you aren't asking for something that a simple Google search would yield. That would be perceived as lazy. It should be obvious that you've done your homework. Multiple posts on the topic will cause people to start associating the idea with you and your personal brand.

For ideas you are very passionate about, an appropriate action would be to write a longer form post. Remind your network that you are thinking about this topic, have gathered some thoughts, and would be interested in theirs. You are building credibility in this area. Doing this through a LinkedIn post or blog post, if you have a blog (or both), can cement this idea with your personal brand, and your network will take notice.

Follow Up Actions

PERSONAL ADVISORY BOARD

Let's also acknowledge that at times during a woman's career, circumstances make it impossible to network in person. This will no doubt be the case at some point in your career. In these times, I've found the need to ramp up a Personal Advisory Board (PAB) to provide some low-stress input and guidance. I did this most recently during the pandemic when I felt starved for opportunities to compare notes with other women as we were all navigating this crazy time. I assembled a group of women who met monthly on Zoom so we could lift each other up. These PAB meetings were a lifesaver for me.

Our current PAB has now supported each other through business pivots, family health crises, and the death of loved ones. We have also done cross-consulting work for each other's companies. This group of four women started out meeting monthly but has since moved to an as-needed basis now that we are back out there in the post-pandemic world. Anyone can call a meeting and drive the agenda, but typically, everyone comes with something they wish to discuss, and we do a round-robin.

A PAB can't replace the need for broader networking events, where you are being exposed to a breadth of new ideas and people. It can be a welcome, friendly alternative and sounding board when that's all that time and energy allow. In an article in *Entrepreneur*, Keith West writes about the emotional support that networking provides, and the PAB was certainly that for us. Keith says, "Many business leaders struggle with feelings of loneliness while running their businesses. By connecting with others, you'll have access to emotional support and camaraderie, which can be invaluable for your mental health and well-being, and you'll also have the opportunity to learn from others and potentially avoid stressful or costly mistakes."[4] Some business struggles require confidential advice from others you trust and can be vulnerable with. This is the perfect application for a PAB.

In our case, PAB support lasts outside of the Zoom calls we hold. We are also each other's fans on social media and share articles and ideas and make introductions to people we think each other should meet. This is all due to the level of intimacy the group has now achieved. The smaller the group, the more intimate they are likely to become.

PIVOT LESSONS YOU CAN USE

1. **Styles vary**. Be aware that men's and women's styles of networking typically vary, so it doesn't throw you off when you encounter different behaviors.

2. **Prep**. Think through what you wish to give and get out of a networking event so you are prepared for those inevitable questions about roles and goals.

3. **Grace**. Networking is easier for some than for others. Finding ways to support those who may find this more difficult is a great gift you can give. Consider other less-tangible things you can give to fellow networkers.

4. **Follow-up.** Actions depend on an assessment of your level of interest in a topic or person. Make the type of follow-up commensurate with the connection you made and the idea it may have inspired.

5. **Don't wait.** Reach out quickly after attending a networking event. Participants will be more open to making professional connections and more likely to remember you.

6. **Event culture.** Every networking event has a culture, and it helps to know what you are walking into so your approach and expectations match the moment.

7. **PAB.** A Personal Advisory Board is a great option for when other networking options aren't possible. They can also be an intimate sounding board for career challenges.

Networking is an essential part of any pivot, but it can be more difficult for some than others and more available to us at certain times in our careers. I learned another valuable lesson about networking from a colleague who recently passed away. When we gathered to appreciate him, one of the common denominators mentioned by coworkers was his practice of calling people just to say hello.

We were each surprised to learn how widespread his calls were. Given how many of us received them, he must have allocated some time each day to this practice. When I received the first of his phone calls, I remember being caught off guard and even suspicious of his real agenda. He let me know that he was calling just to say hello and wondering how my day was going. I came to learn after the next couple of calls that he really was just spending a few minutes checking in.

The effect was a stress-reducing few minutes to connect on work and family topics. It was something I really hadn't experienced before. While I do this with a small circle of friends and family, I haven't done this as a practice with random colleagues. He learned it from an influential mentor of his and adopted it. This is another networking strategy worth mentioning that is appropriate for those people inside your existing network. It's also an act of kindness that is deeply memorable.

9.

What Companies Need to Know

When I told a male relative that I was writing this book, he asked me whether I was planning to talk with any men about the idea of working women navigating careers and family. I furrowed my brow because he seemed to have missed the point. *I really have to work on my book description,* I thought.

This book attempts to raise awareness of the female perspective that seems to go underappreciated as we take on big careers and raise families. It helps women to see each other and encourage support for each other to bring on change. Like my male relative, companies need a female perspective as they define their corporate culture, hire and downsize, create policies, standardize compensation, and establish priorities so they are inclusive of women.

PERCEPTION GAPS AND TECH NEEDS

Globally, companies are kidding themselves about whether their culture is friendly to women. According to an Accenture study of tech companies, 45 percent of HR leaders think it's easy for women to thrive at their organization versus the 21 percent of female employees who do. That gap means we shouldn't be surprised that the proportion of women to men in tech has declined in the last thirty-five years. Women are leaving tech roles at a rate 45 percent higher than men. Fifty percent of women who do take on a tech role drop it by the age of thirty-five, compared with 20 percent of those who do so in other types of jobs.[1] The reasons women are leaving have begun to be studied, and you may be surprised at early findings.

Given that tech roles are a key driver of the US economy, with 86 percent of CIOs saying they face more competition for qualified candidates, qualified women leaving tech roles raises big concerns.[2] Although every company is fast becoming a technology enterprise, Accenture also studied men and women in a variety of industries and found an even broader gap between the perceptions of leaders who feel they create empowering environments (68 percent) and employees who do (36 percent).[3]

Cyberattacks are the primary reason that all companies are fast becoming technology enterprises and, therefore, need to be concerned about qualified women in these roles. As an example, this week, I received a notification from a medical services provider who removed our daughter's wisdom teeth that a breach in their systems had occurred. This put my financial and identification information at risk.

Cybersecurity issues have made it into boardrooms with companies reporting an increase in legal, economic, and internal discussions around cybersecurity. Companies also report an 82 percent increase in their cybersecurity budgets.[4] Government agencies are also struggling with this challenge and incurred $13.7 billion in costs related to cyberattacks in 2018.[5] Cybersecurity alone requires a huge pool of talented resources to service the growing need affecting all companies.

Also concerning is that in 2019, women made up 48 percent of all workers but only 27 percent of STEM workers. STEM workers are broadly defined as social scientists, mathematical workers, life and physical scientists, computer workers, and engineers. Yet while women made gains in the percentage of STEM workers, "women did not make significant gains in computer and engineering occupations, making up the largest portion (80 percent) of the STEM workforce."[6] Both US states and private companies are putting programs in place to educate or reskill the workforce to make sure technical needs can be met. This brings attention to companies supporting women as part of the solution.

To summarize the broader issues raised by these studies, tech skills are highly sought after, women are not choosing proficiency in the tech skills the market is seeking, and the women who do are dropping out of their tech careers in their prime years. Simultaneously, there is a disconnect between whether companies feel they are supporting women in the workplace. That all adds up to a missed opportunity for leveraging capable women who could close the talent gap.

IMPACTING STEREOTYPES

Let's start with why women aren't choosing computer science and engineering fields. *Scientific American* believes it has to do with early stereotypes that signal that girls aren't interested in these fields. That, combined with male-oriented images and masculine defaults such as aggressiveness, overconfidence, and self-promotion, tends to push girls away at early ages. Most people's stereotypical impressions of a computer scientist are driven by media like *Big Bang Theory* or *Silicon Valley*, which predominantly feature "mostly white, sometimes Asian male geniuses who are socially awkward, play video games and like science fiction."[7]

Unfortunately, once pushed away, some girls never rediscover their interest in STEM. This is reflected in data showing that only 28 percent of students taking high school Advanced Placement computer science exams are women. The women who do make it as far as college programs experience a higher switch rate (37 percent) than other majors.

The Accenture study suggests this switch rate is related to women not finding enough people like themselves in these high school and college programs, causing a feeling of isolation. Women "do not thrive on par with men in less inclusive cultures." Women who study in a more inclusive college culture are more likely to ask questions, feel they belong, and feel inspired by their classes and then consider looking for jobs in tech.[8]

A very wise woman once told me that her secret to completing her chemistry degree was finding her "math friends." She made a concerted effort to establish a group of other women to study and collaborate with. This was a very insightful approach that the Accenture research now validates. Women need to be surrounded with other women they relate to and who support them in their endeavors. Of course, this is true of men too. It's just that men have many more "math friends" to begin with.

Being aware of these realities and devising long-term strategies that encourage women at the high school and college levels to consider STEM careers is important for both companies and parents. For example, *Big Four* accounting and technology firms are now sending a diverse team of representatives to high school career fairs. They also send a diverse team of young associates to lead STEM programs in minority neighborhoods. NASA sponsors and mentors at high school robotics competitions. While these are important efforts, high school is too late to catch girls exploring interest in these topics, as discussed in chapter 7.

Keeping women in their tech roles once they have arrived should be a focal point for companies. The solutions around an inclusive workplace culture are certainly relevant here. Accenture suggests the number of women in tech could potentially be doubled by 2030 if every workplace was on par with the top 20 percent of those in their study.[9]

Companies can start by measuring the earlier mentioned gap between beliefs that their leaders are creating an inclusive environment and the reality of female employees' experience to understand how much work needs to be done. Grasping how women feel about their job, whether they have been promoted or expect to be, aren't mistaken for being more junior than male peers, aren't made to feel like the job isn't for "people like them," nor have heard or read inappropriate comments, all correlated to how likely women are to leave their chosen company or field. Of course, measuring this gap across all workers would be necessary to capture differences in perceptions from women of color, LGBTQ+, or other minority groups in your organization.[10]

Adjusting for masculine defaults in job descriptions would be an important step to aligning corporate culture. It may also be necessary to call on qualified women to consider tech job openings, as Google has reported learning. After realizing that women weren't applying for tech promotions at the same rate as men, Google looked at two studies on gender inequality for insights into what could be causing this. They hypothesized that the problem was the volunteer system.

The studies told them that "Girls don't raise their hands as often as boys when answering math problems, even though they have a higher accuracy rate when they do. Women don't offer up their ideas as often as men in business meetings, even though observers say their thoughts are often better than the many offered by their male colleagues." Google made a point of sharing this data with employees and encouraging all to apply. When slots opened for engineering positions "immediately, the application rate for women soared, and the rate of women who received promotions rose higher than that for male engineers."[11]

Throughout this book, I have talked about how and why women operate differently than men, but this Google experiment illustrates the ambition of women once they realize the company is interested in their success and progression. Once cultural barriers have been removed and the dialogue is open, women can thrive.

Accenture identified a total of forty cultural factors that impact the retention of women and other employees. They can be grouped into three buckets. "Bold Leadership" refers to a diverse team that openly sets, shares, and measures targets. "Comprehensive Action" includes family friendly policies and practices that are bias-free when attracting and retaining people. "Empowering environments" breed trust and respect for employees, offering freedom to be creative, train and work flex.[12]

INNOVATION AND PROFITABILITY

With these three factors in mind, a culture of inclusion quickly becomes a culture of innovation when employees are less afraid to fail and see fewer barriers to innovating. Accenture goes on to unpack the impact on profitability when employees have an "innovation mindset." Six elements of an innovation mindset were identified including purpose, resources, collaboration, autonomy, inspiration, and experimentation.

Accenture found that "innovation mindset is six times higher in the most-equal cultures than in the least-equal ones." They also estimate that "global gross domestic product would increase by up to $8 trillion by 2028 if innovation mindset in all countries were raised by 10 percent."

A final key finding of this study suggests that "increasing pay is considerably less effective than bolstering a more equal culture,"[13] so we can't throw money at innovation. Corporate culture must be changed to reflect the inclusion of ideas and participation.

COMPENSATION EQUALITY

The prior point about not buying innovation should not be confused with equality in compensation. According to the Center for American Progress, women continue to be paid less than men. CAP broke down the compensation gaps between men and women by age range, revealing that the gaps worsen as women age: Women eighteen to twenty-four earn 8 percent less than men; women twenty-five to fifty-four earn 16 percent less; women fifty-five to sixty-four earn 22 percent less; and women over sixty-five earn 27 percent less. Earning gaps are even larger for women of color.

The report goes on to say that "having minor children at home still disproportionately depresses women's employment prospects. The effect is largest for those with the youngest children—and this was the case before the COVID-19 recession." Given that women are five to eight times more likely to have their employment affected by caregiver responsibilities, the growing compensation gap you see in later years suggests they never really recover. "Time out of the labor force and reduced working hours have lifelong ramifications for women's economic security—from lost earnings today to smaller Social Security benefits and retirement savings down the road."[14]

State legislatures are now adopting equal pay laws to combat this. "Most state laws provide broader protection by requiring employers to pay men and women equally for 'substantially similar' work, rather than for *equal* work." With the passing of new legislation comes litigation against employers, so SHRM recommends that companies conduct pay audits. "Through the audit, an employer can determine if discrepancies can be explained by legitimate, nondiscriminatory reasons." If they can't be, then companies have their list of issues worth resolving. Also recommended are things like "controls for salaries offered at the hiring stage," which prevent the temptation to offer higher compensation packages to new hires during a tight labor market. Once companies understand and rightsize their compensation, it boils down to being transparent about "systems and objective metrics around recruitment, performance, advancement, and compensation to help ensure consistency."[15] This followed by regularly communicating with employees about these issues builds trust, which fosters a strong company culture.

LEADERSHIP OPPORTUNITIES

Like the issue of compensation disparity being a chronic problem for women, so is their representation at the leadership level. "Even in 2023, women still face challenges to their authority and success that are greater than those faced by their male counterparts. Women are slowly rising in political leadership and in corporate and educational leadership."[16] While the numbers slowly trend upward and momentum builds, the differences in leadership styles between men and women were labeled in the Eagly study in 2013.

The study refers to the "transactional" and more male style of leadership versus the "transformation" and more female style. While Eagly acknowledges that the differences are small, they lie in values and attitudes of men and women. "One of these differences is that female leaders, on average, are more democratic and participative than their male counterparts. Men, more than women, adopt a top-down, 'command and control' style."[17]

It's harder to draw a straight line between female leadership and profit (although some studies do), it's easier to draw that line to collaboration. "The proportion of women in a group was strongly related to the group's collective intelligence, which is their ability to work together and solve a wide range of problems. Groups with more women exhibited greater equality in conversational turn-taking, further enabling the group members to be responsive to one another and to make the best use of the knowledge and skills of members."[18]

Let's tie together a few different threads. First, depending on corporate culture, women may need to be drawn out as participants and leaders, as proven by the Google experiment mentioned earlier. The second is that women leaders are more likely to be the ones to introduce "democratic" style into a company's culture. Third, the unique leadership characteristics that women are most comfortable using are likely to work well in companies striving for innovation. But most importantly, the pipeline needs to be full of a diverse pool of potential leaders to have the right leader to choose from when openings come. It is important to make sure women are included at all levels of management—especially in the C-suite, which has proven to get the diversity ball rolling quickly.

Improving this pipeline of women in leadership roles leads to better representation of women on corporate boards. According to the most recent studies, women held "27 percent of board seats at companies on the Russell 3,000 in 2021, up from 24 percent in 2020." There is a ripple effect associated with women in leadership roles. "The average percentage of women on boards of companies with a female CEO is 39 percent, while it is 26 percent for companies with a male CEO."[19]

So women seeking board roles know to look for them at companies where there is already a strong pool of women leaders and, therefore, a devotion to diversity. Assigning one woman to a senior leadership role is often the first step in diversifying an organization.

MENTORSHIP, SPONSORSHIP, AND ALLYSHIP

Another important step is through creating spaces where women can discuss their questions and concerns openly. *Monster* talks about the concepts of "mentorship, sponsorship, and allyship." Mentorship programs can be both formal or informal, using a mentor/protégé arrangement that pairs a senior leader with a more junior or new woman in the organization. In my experience, mentoring programs are most successful when some accountability is built in, such as meeting frequency and some key performance metrics (KPI) for both the mentor and the protégé.[20]

Sponsorship programs differ in connecting "someone with the power to further your career, who is convinced of your potential and can speak on your behalf when promotion decisions are made."[21] In this case, sponsors become a part of the promotion process and function as advocates who know the person's goals and objectives. These programs are most successful when they are clearly defined and available to all employees—not just a chosen few. The advocacy component helps companies identify talent that might get overlooked and support diversity goals.

Allyship is more of a peer-to-peer support structure, which can be formalize through Employee Resource Groups (ERG), or through encouraging safe spaces for discussion. "Allyship takes place at the coworker level. An ally might be someone who women can feel safe sharing frustrations with, especially frustrations that are a consequence of gender bias."[22] In my experience, creating allyship programs require HR involvement to ensure quality and consistency while providing guidance on how peer-to-peer support should, and shouldn't, work. Baking guidelines into new employee orientations and then reinforcing them during HR check-ins can help make sure that a framework is being followed.

All three of these programs help companies keep their finger on the pulse of employee satisfaction and open a wider dialogue.

PIVOT LESSONS COMPANIES SHOULD CONSIDER

1. **They just leave.** Women won't tell you why they are leaving. They just do it. Unless companies are prepared to ask about job satisfaction, they won't see it coming until they have lost a valuable asset.

2. **Support early.** More companies are finding ways to support young women through internships and STEM program sponsorship that normalize STEM for women and girls. Consider ways your company can influence girls' perceptions—even at the elementary school level.

3. **Mind the gap.** Many companies experience gaps in how well they think they understand the issues facing women in the workplace. Measure to understand your gap and institute ways to close it.

4. **Diversity breeds innovation and profitability.** This gets measured in different ways, but more companies, especially innovative ones, are seeing strong reasons to consider diversity a priority at all levels of the organization. Recruiting women in senior leadership roles is proven to be a strong step in getting the ball rolling.

5. **Audit compensation.** Audit and rightsize compensation throughout your organization to send a strong signal that diversity is serious business.

6. **Mentorship, sponsorship, allyship.** Find the right combination of these types of programs based on your company size and goals.

Women are key to the success of organizations that face continued technological and business challenges. However, they will need to feel encouraged and supported to stay engaged as they juggle their unique challenges of work/life balance talked about throughout this book.

Conclusion

As I said at the outset, I was under the false impression that a career yields a consistently upward trajectory, meaning making more money each year and getting a higher-ranking title with each new role. Reality and the research I've done for this book show that is often not the case for women.

Too many interruptions are unique to us, and recovery can be slow with long-term implications. Knowing this has changed my understanding of success and reminds me not to compare myself to others—especially to men. That comparison cannot be made for all the reasons outlined in this book. Women will likely have to make many trade-offs in their careers. That is not failure.

At the beginning of this book, I mentioned I was purging some demons. I went back and looked at family and career situations from the standpoint of what I learned and what I would do differently with the benefit of hindsight. In analyzing these situations, two things happened. One, I was able to be more objective about the complexity of the situations, and I could more easily see multiple sides. Two, through reading a ton of research on the dynamics of women's careers, I was able to forgive myself for not navigating these situations more forcefully, shrewdly, eloquently, or creatively. I highlighted the lessons I learned throughout this book, and I no longer second-guess the decisions I made.

In doing my research, the topic that continued to intrigue me was about how women are perceived. They must be more subtle in their delivery. From a young age, people have told me I should smile more. Moving from the Northeast to the Mid-Atlantic has helped me tone down my "don't-mess-with-me" vibe, but even as I've recently welcomed a new son-in-law and daughter-in-law into our family, they have shared stories of how intimidating I can come across at a first meeting.

Realizing I still have more toning down to do, I took special interest in the materials I was reading on how women aren't as successful in networking, leading, or self-promoting when they do them in the same way as men. We must be less "on-the-nose" to be effective—more nuanced. No doubt, the advice I received to smile was getting at this. They were telling me to be softer, which I can now appreciate in a rational way.

The self-promotion skill will always be the hardest for me and for many of the women I spoke to. Women my age acknowledge that it's not something we were trained to do, so it doesn't come naturally to many of us. I'm currently practicing this skill every chance I get and trying to compliment other women when I see them do it well.

The topic of mentors and role models also interested me, and I'm finishing this book feeling more appreciative of the informal mentors I've had. Everyone has a role model that you can admire from afar or use as a counter example, but not everyone has had a formal mentor. Those with extensive formal educations have likely had them because mentors are built into that experience. But if you haven't participated in a formal mentoring program through school, work, or sports, you may feel as though you missed out. I did.

Given the broader, informal definition of a mentor, we've all had someone who took a special interest and probably didn't realize how much we appreciated their support. I was fortunate to know Rick Blair at AOL, which was a time when I had a big, ever-changing professional role, and young children. As the father of two children himself, he gave trustworthy advice while also providing insightful, diplomatic observations that I greatly appreciated then and still remember now. He never told me to smile more.

Anyone in the throes of raising young children and still finding time to read this book, I send you props. The intensity of this phase does wane, which is hard to imagine when you are in the thick of it. I encourage you to be patient with yourself as you navigate challenges during this time. In the chapter "The Business of Mothering," I talked about the importance of a support network and prepared readers for the ways your needs will change as your children grow.

The old saying "little kids, little problems—big kids, big problems" seems relevant here. Although I would modify it to "little kids, little problems—big kids, lots of driving, drama, and uncertainty." It really does take a village, and you'll have to assemble and reassemble your village as your needs evolve. Sometimes, the support network feels very tenuous, but the larger your village, the more manageable it all becomes.

Others are assembling their village, too, so plug in where you can. This will be important to your survival. My wise friend and colleague Donita Prakash, CEO of Brand Transitions, told me that "my advice to my younger self would be don't put so much pressure on yourself. It all works out. But at the same time, you need to prioritize personal stuff, too."

At the beginning of this book-writing process, I was diving into what made some women more willing to take on big roles—to be adventurous. Most women I spoke to said they had great encouragement from friends and family, but they also had challenges to navigate. These women were highly qualified, which isn't surprising since most would only consider taking on roles they were 100 percent qualified for. The biggest common characteristic between them was that they identified a big goal and ran at it. Sometimes, they second-guessed themselves along the way, but then they just kept chipping away. It's so important to see what you can be, so get out there and be an observer.

Big goals sometimes change, which was the case for me after my iBelong experience. It was no longer important for me to run a company with lots of investors and partners who brought multiple agendas and levels of influence. I saw too many trade-offs would be necessary, and I wasn't willing to make them.

Goals for control and flexibility are common for many women, as proven by the data cited in chapter 5: "Entrepreneurialism." We also now have data showing that women are great at enabling innovation, as cited in chapter 9: "What Companies Need to Know." That means we can expect more Annie Wojcicki-level innovation and Geena Davis-level collaboration in years to come. I'll be excitedly watching this. I'll also take notice as women experiment with disruptive behaviors that don't come naturally.

As many of my female friends are navigating what their "retirement" phase looks like, we take cues from each other. We are curiously watching each other make relationship decisions, try new things, downsize, travel, and become grandparents. For me, I'll keep seeking a combination of the familiar and unfamiliar.

I'll deploy familiar technology, leadership, transformation, and marketing skills to work at the C-level and board level of interesting companies. However, I'm staying open to those unfamiliar opportunities that may now present themselves through my new author community. I've also formalized the mentoring I'm doing with entrepreneurs through George Mason University and the ICAP Program. I can't stay away from entrepreneurs' energy and ideas. I'm especially interested in guiding more women in these roles.

While women continue to make progress in their careers, we need each other's support. We are the best qualified to do this because we understand the playing field so well. I'm hoping after reading this book, you will have identified ways that you can impact your companies through programs, leadership, and culture. Honor yourself by acknowledging the things you have done well and then let others know, too. I'm also hoping you will seek one-on-one opportunities to support women in your communities. Don't forget to ask for and accept the help you may need in the process.

My final thought is that none of us have this all figured out—no matter what our age—nor are we supposed to. Pivoting is never-ending. I'm taking some pressure off myself to be perfect in favor of being kind. Remember to let someone know how they have positively impacted you. Maybe even call someone just to see how their day is going.

Acknowledgments

This book wouldn't have been possible without the patience of those who trusted me with their stories. You generously lent your time, perspectives, and support. Thank you to: Seema Alexander, Nina Brugel, Jenna Close, Irene Hakes, Janet Hall, Reggie Kouba, Carmel McDonagh, Dr. Marian McKee, Nancie Laird Young, Monica Pemberton, Donita Prakash, Kristen Runke, Shelly Ryan, Elaine Tholen. Not everything made it into the book, but please know that you all fortified my thinking.

There were many beta readers of sections of the book, but two people went above and beyond by reading most of it. You saw things in raw form and were able to look past that to provide comments that were so insightful. You kept me in check, asked great questions, and encouraged me so much. Thank you to Nancie Laird Young and Rumy Sen.

I'd also like to thank Stacy Kleber Jensen for her work on the graphics for this book. She is so talented, and her work always makes things clearer and memorable.

These projects don't happen without regular support and encouragement. To all those who were part of my presale campaign, you gave me the courage to keep working at this daunting task. I can't tell you how much that meant at the early stages. Thank you to Jan Ahearn, Maris Angolia, Clayton and Courtney Barber, Fiona Barber, John and Laurie Barber, Maura Barber, Robert and Jane Barber, Haven Barbieri, Leslie Bauer, Richard Blair, Angela Bongiorno, Barbara Breivik, Nina Brugel, Gretchen Buckler, Lily and Bryan Burns, Clara Conti, Beth Cox, Marianne Cullen, Ashley Da Silva, Patricia Daunas, Patricia Debearn, Sue Deighton, Pamela Diganci, Shannon and Joe Doretti, Karen and Michael Fisher, Kristin and Lance Fitzmorris, Stephen Gray, Tatiana Griffin, Burl Haigwood, Kathryn Harris, Teresa Hawkins, Lisa Hodge, Megan Holmes, Stacy Jensen, Lauren Johns, L. Larry Kibler, Joan Kickert, Taylor Kiland, Jim Kovarik, Irene Lane, Elena and Mike Langlois, Elise Logan, Jennifer Lynch, Lisa Malachowsky, Doreen Mannion, Alden Downs and Kim McNeill Downs, Beth McQuaid, Nancy Munyan, Michael Nelson, Carmen Radelat, M'Liz Riechers, Mark H Rooney, Christina Rossi, Jean Sammarco, Valerie Sandlas, Fern Schwartz, Lea Schwarzenberg, Tarun Sen, Jenny Song, Kate Sprague, Bob Smith, Felicity Tagliareni PhD, Ellen Tenenbaum, Amy Tierney, Dr. Eva Vinson, Boni Vinter, Susan and Randy Walker, Grace Walker, Rose Wang, Lindsey Waugh, Kevin Wicks, and Mary Willinger.

And thanks to the amazing team at Manuscripts, who provided guidance and a process to follow. You were patient and taught me so much: Cassandra Caswell-Stirling, Kyra Ann Dawkins, Kristy Elam, Shanna Heath, Angela Ivey, Jill Magnuson, Stephanie McKibben, Tatiana Obey, Katie Sigler,

Chrissy Wolfe. To Gjorgji Pejkovski and his team, who did the cover design, and to Eric Koester as the grand master of inspiration, thank you.

Appendix

CHAPTER 1—CAREER GROWTH DURING LIFE PHASES

1. David A. Morin, "Social Life Struggles of Women in their 20s and 30s," *Relationships* (blog), *SocialSelf*, August 4, 2022, https://socialself.com/blog/social-life-struggles-women/.

2. Ibid.

3. Ibid.

4. Sylvia Ann Hewlett, "Executive Women and the Myth of Having It All," *Harvard Business Review*, April 2002. https://hbr.org/2002/04/executive-women-and-the-myth-of-having-it-all.

5. Aspen Institute, "Henry Crown Fellowship," Henry Crown Fellowship, Aspen Institute, September 26, 2023, https://www.aspeninstitute.org/programs/henry-crown-fellowship/.

6. Annette Joan Thomas, Ellen Sullivan Mitchell, and Nancy Fugate Woods, "The Challenges of Midlife Women: Themes from the Seattle Midlife Women's Health Study," *Women's*

Midlife Health 4, no. 8 (2018): 4, https://doi.org/10.1186/s40695-018-0039-9.

7. Sylvia Ann Hewlett, "Executive Women and the Myth of Having It All," *Harvard Business Review*, April 2002, https://hbr.org/2002/04/executive-women-and-the-myth-of-having-it-all.

8. Russell Heimlich, "Number of Americans Who Read Print Newspapers Continues to Decline," *Research Topics* (blog), *Pew Research Center*, October 11, 2012, https://www.pewresearch.org/short-reads/2012/10/11/number-of-americans-who-read-print-newspapers-continues-decline/.

9. Paul Taylor, Cary Funk, and Peyton Craighill, *Working After Retirement: The Gap Between Expectations and Reality* (Washington DC: Pew Research Center, 2006), 2, https://www.pewresearch.org/social-trends/2006/09/21/working-after-retirement-the-gap-between-expectations-and-reality/.

10. Lauren Aratani, "Goodbye to the Job. How the Pandemic Changed Americans' Attitude to Work," *Money* (blog), *The Guardian*, November 28, 2021, https://www.theguardian.com/money/2021/nov/28/goodbye-to-job-how-the-pandemic-changed-americans-attitude-to-work.

CHAPTER 2—WHY DO WE PIVOT?

1. Morgan Smith, "GenZ and Millennials Are Leading 'The Big Quit' in 2023—Why Nearly 70% Plan to Leave Their Jobs," *Make It* (blog), *CNBC*, January 18, 2023, https://www.cnbc.com/2023/01/18/70percent-of-gen-z-and-millennials-are-considering-leaving-their-jobs-soon.html.

2. Katherine Weisshaar, "From Opt Out to Blocked Out: The Challenges of Labor Market Re-entry After Family-Related Employment Lapses," *American Sociological Review* 83, no. 1 (January 10, 2018), 44 https://doi.org/10.1177/0003122417752355.

3. Dr. Gleb Tsipursky, "Women's Day: The Vital Importance of Allowing Women to Have a Flexible Hybrid Schedule," *Leadership Strategy* (blog), *Forbes*, March 8, 2023, https://www.forbes.com/sites/glebtsipursky/2023/03/08/womens-day-the-vital-importance-of-allowing-women-to-have-a-flexible-hybrid-schedule/.

4. Bernard Schroeder, "Gen Z Evaluates Going to College or Getting Important Job-Based Certifications. Why Certain Careers Might be Better Suited to Certifications," *Small Business Strategy* (blog), *Forbes*, December 2, 2019, https://www.forbes.com/sites/bernhardschroeder/2019/12/02/gen-z-evaluates-going-to-college-or-getting-important-job-based-certifications—why-certain-careers-might-be-better-suited-to-certifications/?sh=45ea55582911.

5. Indeed Editorial Team, "10 In-Demand Career Certifications (And How to Achieve Them)," *Career Development* (blog), *Indeed*, October 3, 2023, https://www.indeed.com/career-advice/career-development/certifications-in-demand.

6. Caroline Castrillon, "Why Women Leaders Are Leaving Their Jobs at Record Rates," *Careers* (blog), *Forbes*, May 7, 2023, https://www.forbes.com/sites/carolinecastrillon/2023/05/07/why-women-leaders-are-leaving-their-jobs-at-record-rates/.

7. Amanda Barroso, "For American Couples, Gender Gaps in Sharing Household Responsibilities Persist Amid Pandemic," *Research Topics/Coronavirus (COVID-19)* (blog), *Pew Research Center*, January 25, 2021, https://www.pewresearch.org/short-reads/2021/01/25/for-american-couples-gender-gaps-in-sharing-household-responsibilities-persist-amid-pandemic/.

8. Kiara Alfonseca, "DEI: What Does It Mean and What Is Its Purpose," *ABC News*, February 10, 2023, https://abcnews.go.com/US/dei-programs/story?id=97004455.

9. Rachel Minkin, "Diversity, Equity and Inclusion in the Workplace," *Business & Workplace* (blog), *Pew Research Center*, May 17, 2023, https://www.pewresearch.org/social-trends/2023/05/17/diversity-equity-and-inclusion-in-the-workplace/.

10. Rakesh Kochhhar, "The Enduring Grip of the Gender Pay Gap," *Economics, Work & Gender* (blog), *Pew Research Center*, March 1, 2023, https://www.pewresearch.org/social-trends/2023/03/01/the-enduring-grip-of-the-gender-pay-gap/.

11. Stephanie Ferguson and Isabella Lucy, "Data Deep Dive: A Decline of Women in the Workforce," *Workforce* (blog), *US Chamber of Commerce*, April 27, 2022, https://www.uschamber.com/workforce/data-deep-dive-a-decline-of-women-in-the-workforce.

12. World Health Organization Editorial Team, "COVID-19 Pandemic Triggers 25% Increase in Prevalence of Anxiety and Depression Worldwide," *News* (blog), *World Health Organization*, March 2, 2022, https://www.who.int/news/

item/02-03-2022-covid-19-pandemic-triggers-25-increase-in-prevalence-of-anxiety-and-depression-worldwide.

13. Amy Novotney, "Women Leaders Make Work Better. Here's the Science Behind How to Promote them," *Women and Girls* (blog), *American Psychological Association,* March 23, 2023, https://www.apa.org/topics/women-girls/female-leaders-make-work-better.

14. Meredith Somers, "Women Are Less Likely than Men to Be Promoted. Here's One Reason Why," *Gender* (blog), *MIT Management Sloan School,* April 12, 2022, https://mitsloan.mit.edu/ideas-made-to-matter/women-are-less-likely-men-to-be-promoted-heres-one-reason-why.

15. Whitney Johnson and Tara Mohr, "Women Need to Realize Work Isn't School," *Gender* (blog), *Harvard Business Review,* https://hbr.org/2013/01/women-need-to-realize-work-isnt-schol.

16. Ibid.

17. Ibid.

18. Kat Tretina and Taylor Tepper, "The Average Age of Retirement in the US," *Retirement* (blog), *Forbes Advisor,* Oct 13, 2022, https://www.forbes.com/advisor/retirement/average-retirement-age/.

19. Brittany King, "Those Who Married Once More Likely Than Others to Have Retirement Savings," *America Counts: Stories* (blog), United States Census Bureau, January 13, 2022, https://

www.census.gov/library/stories/2022/01/women-more-likely-than-men-to-have-no-retirement-savings.html.

CHAPTER 3—NAVIGATING ORGANIZATIONAL CHAOS

1. Yekaterina Chzhen, Anna Gromada, Gwyther Rees. *Are the World's Richest Countries Family Friendly?* (Florence, Italy, Unicef Office of Research—Innocenti, 2019) 6, https://www.unicef-irc.org/publications/pdf/Family-Friendly-Policies-Research_UNICEF_%202019.pdf.

2. Arden Davidson. "How to Sniff Out a Great Company Culture During Your Job Search," Jobs (blog), The Washington Post, March 17, 2022, https://jobs.washingtonpost.com/article/how-to-sniff-out-a-great-company-culture-during-your-job-search/.

CHAPTER 4—IDENTITY

1. Jan Stets and Peter Burke, "Self-Esteem and Identities," *Sociological Perspectives* Volume 57, Issue 4 (December 2014):1,2, https://doi.org/10.1177/0731121414536141.

2. Kelly Corrigan, "Mistakes or The Only Way? With Michelle Icard," *Kelly Corrigan Wonders*, August 22, 2023, 00:52:53, https://www.kellycorrigan.com/kelly-corrigan-wonders/michelleicard.

3. John Heilemann, "Jennifer Psaki," *Hell & High Water with John Heilemann*, October 4, 2022, 01:20:19, https://shows.acast.com/612830d1407ad60013fa691a/episodes/jen-psaki.

4. Jennifer Liu, "The Best Career Advice White House Press Secretary Jen Psaki Ever Got," *Make It* (blog), *CNBC*, April 1, 2021, https://www.cnbc.com/2021/04/01/the-best-career-advice-white-house-press-secretary-jen-psaki-ever-got.html.

5. John Heilemann, "Jennifer Psaki," *Hell & High Water with John Heilemann*, October 4, 2022, 01:20:19, https://shows.acast.com/612830d1407ad60013fa691a/episodes/jen-psaki.

6. Kara Swisher, "Jen Psaki on Her New Show and Her Old Boss," *On with Kara Swisher*, April 17, 2023, 00:54:26, https://open.spotify.com/episode/4Tfb4Ngwvd7nOoaEWj93QP.

CHAPTER 5—ENTREPRENEURIALISM

1. Kelsey Miller, "10 Characteristics of Successful Entrepreneurs," *Business Insights* (blog), *Harvard Business School Online*, July 7, 2020, https://online.hbs.edu/blog/post/characteristics-of-successful-entrepreneurs.

2. Chris Kuenne and John Danner, *Built for Growth: How Builder Personality Shapes Your Business, Your Team, and Your Ability to Win* (Boston, Massachusetts: Harvard Business Review Press, 2017), Cover Blurb.

3. Chris Kuenne and John Danner, *Built for Growth: How Builder Personality Shapes Your Business, Your Team, and Your Ability to Win* (Boston, Massachusetts: Harvard Business Review Press, 2017), 86–87, 161

4. Chris Kuenne and John Danner, *Built for Growth: How Builder Personality Shapes Your Business, Your Team, and Your Abil-

ity to Win (Boston, Massachusetts: Harvard Business Review Press, 2017), 2.

5. Steve Blank and Bob Dorf, *The Startup Owner's Manual* (Hoboken, New Jersey: John Wiley & Sons Inc., 2020), xvii–xviii.

6. Jessica Elliott, "Women in Small Business Statistics in the US," *Small Business* (blog), *The Ascent*, August 5, 2022, https://www.fool.com/the-ascent/small-business/articles/women-in-small-business-statistics-in-the-us/.

7. Rieva Lesonsky, "The State of Women Entrepreneurs," *SCORE* (blog), *SCORE,* February 6, 2023, https://www.score.org/resource/blog-post/state-women-entrepreneurs.

8. Timothy Butler, "Hiring and Entrepreneurial Leader," *Harvard Business Review,* March–April 2017 https://hbr.org/2017/03/hiring-an-entrepreneurial-leader.

9. Sarah Zhang, "Big Pharma Would Like Your DNA," *Science* (blog) *The Atlantic*, July 27, 2018, https://www.theatlantic.com/science/archive/2018/07/big-pharma-dna/566240/.

10. GSK Global, "GSK and 23andMe Sign Agreement to Leverage Genetic Insights for the Development of Novel Medicines, GSK Media/Press Releases, GSK, July 25, 2018, https://www.gsk.com/en-gb/media/press-releases/gsk-and-23andme-sign-agreement-to-leverage-genetic-insights-for-the-development-of-novel-medicines/.

11. Crunchbase (object name 23andMe; accessed August 25, 2023), https://www.crunchbase.com/organization/23andme/company_financials.

12. Rohit Arora, "Women-Owned Businesses Thrived in 2022," *Small Business* (blog), *Forbes*, March 8, 2023, https://www.forbes.com/sites/rohitarora/2023/03/08/women-owned-businesses-thrived-in-2022/.

13. Timothy Carter, "The True Failure Rate of Small Businesses," *Starting a Business* (blog), *Entrepreneur*, January 3, 2021, https://www.entrepreneur.com/starting-a-business/the-true-failure-rate-of-small-businesses/361350.

CHAPTER 6—THE BUSINESS OF MOTHERING

1. Parker, Juliana, Menasce Horowitz, and Milly Rohal, "Raising Kids and Running a Household: How Working Parents Share the Load," (Washington, DC: Pew Research Center, 2015) 3, https://www.pewresearch.org/social-trends/2015/11/04/raising-kids-and-running-a-household-how-working-parents-share-the-load/.

2. Aria Bendix and Joe Murphy, "The Modern Family Size Is Changing. Four Charts Show How," *Data Graphics* (blog), *ABC News*, January 12, 2023, https://www.nbcnews.com/health/parenting/how-modern-us-family-size-changing-charts-map-rcna65421.

3. BLS Beta Labs (BLS Data Viewer/Day care and preschool in US city average, all urban consumers, seasonally adjusted;

accessed September 3, 2023), https://beta.bls.gov/dataViewer/view/timeseries/CUSR0000SEEB03.

4. Kiley Hurst, "More than Half of Americans Live within an Hour of Extended Family," *Family & Relationships* (blog), *Pew Research Center*, May 18, 2022, https://www.pewresearch.org/short-reads/2022/05/18/more-than-half-of-americans-live-within-an-hour-of-extended-family/.

5. Wyndi Kappes, "6 Reasons Why Women's Labor Participation Just Hit a 33-Year Low," *News* (blog), *The BUMP*, June 10, 2022, https://www.thebump.com/news/record-low-womens-labor-participation.

6. Erin Rehel and Emily Baxter, *Men, Fathers, and Work-Family Balance* (Washington, DC: Center for American Progress, 2015), https://www.americanprogress.org/article/men-fathers-and-work-family-balance/.

7. Lorin Cox, "Why More Men Are Leaving the Workforce to Care for Children and Families," January 3, 2023, Wisconsin, Onair Recording, 0:21:30, https://www.wpr.org/listen/2042731.

8. Heather Gillin, "How Career Interruptions Affect Women," *Culture & Society* (blog), *Texas A&M Today*, March 9, 2020, https://today.tamu.edu/2020/03/09/how-career-interruptions-affect-women/.

CHAPTER 7—UNEXPECTED WAYS OF IMPACTING PERSONAL BRAND

1. Christine Comaford, "76% of People Think Mentors Are Important, but Only 37% Have One," *Leadership Strategy* (blog), *Forbes*, July 3, 2019, https://www.forbes.com/sites/christinecomaford/2019/07/03/new-study-76-of-people-think-mentors-are-important-but-only-37-have-one/?sh=2580abb04329.

2. W. Brad Johnson, David G. Smith, and Jennifer Haythornthwaite, "Why Your Mentorship Program Isn't Working," *Business Management* (blog), *Harvard Business Review*, July 17, 2020, https://hbr.org/2020/07/why-your-mentorship-program-isnt-working.

3. Suzanne de Janasz and Maury Peiperl, "CEOs Need Mentors Too," *Harvard Business Review*, April 2015, https://hbr.org/2015/04/ceos-need-mentors-too.

4. Knowledge at Wharton Staff, "Workplace Loyalties Change, but the Value of Mentoring Doesn't," *Knowledge at Wharton Podcast*, released May 16, 2017, 0:16:26, https://knowledge.wharton.upenn.edu/podcast/knowledge-at-wharton-podcast/workplace-loyalties-change-but-the-value-of-mentoring-doesnt/.

5. Lisa Quast, "How Becoming a Mentor Can Boost Your Career," *Leadership* (blog), *Forbes*, October 31, 2011, https://www.forbes.com/sites/lisaquast/2011/10/31/how-becoming-a-mentor-can-boost-your-career/.

6. Lisa Quast, "Why Women Should Have Career Mentors," *Leadership* (blog), *Forbes*, July 18, 2011, https://www.forbes.com/

sites/lisaquast/2011/07/18/why-women-should-have-career-mentors/?sh=44d5ff1c1813.

7. Claude Balthazard, "#1 the Four Types of Professional Development Organizations," *Pulse* (blog), *LinkedIn*, April 4, 2017, https://www.linkedin.com/pulse/four-types-professional-organizations-claude/.

8. Kara Swisher, "Are You 'Dying of Politeness'? Geena Davis Explains," *On with Kara Swisher*, Jan 9, 2023, 00:46:08, https://open.spotify.com/episode/2h0RV7WFk4C5RZ40kZEc6W.

9. Geena Davis, *Dying of Politeness* (New York, NY: HarperCollins Publishers, 2022), 2.

10. Lin Bian, Sarah-Jane Leslie, and Andrei Cimpian, "Gender Stereotypes about Intellectual Ability Emerge Early and Influence Children's Interests," *Science* 355, no. 6323 (January 2017): 389–391, https://www.science.org/doi/10.1126/science.aah6524.

11. Werner Geyser, "8 Personal Brand Statement Examples to Help You Craft Your Own Brand," *Influencer Marketing* (blog), *Influencer Marketing Hub*, July 15, 2022, https://influencermarketinghub.com/personal-brand-statement-examples/.

12. Geena Davis, *Dying of Politeness* (New York, NY: HarperCollins Publishers, 2022), 254.

13. Geena Davis Institute on Gender in Media, "What's the Issue?" About Us, Geena Davis Institute on Gender in Media, accessed September 29, 2023, https://seejane.org/about-us/.

14. Geena Davis, *Dying of Politeness* (New York, NY: HarperCollins Publishers, 2022), 276.

CHAPTER 8—NETWORKING

1. Caroline Castrillon, "Why Women Need to Network Differently than Men to Get Ahead," *Careers* (blog), *Forbes*, March 10, 2019, https://www.forbes.com/sites/carolinecastrillon/2019/03/10/why-women-need-to-network-differently-than-men-to-get-ahead/?sh=1a3f52ebb0a1.

2. Ibid.

3. Frencesca Gino, Maryam Kouchaki, and Tiziana Casciaro, "Learn to Love Networking," *Harvard Business Review*, May 2016, https://hbr.org/2016/05/learn-to-love-networking.

4. Keith West, "5 Surprising Benefits of Professional Networking That You Need to Know About," *Leadership* (blog), *Entrepreneur*, April 17, 2023, https://www.entrepreneur.com/leadership/5-surprising-benefits-of-professional-networking/448862.

CHAPTER 9—WHAT COMPANIES NEED TO KNOW

1. Accenture and Girls Who Code, *Resetting Tech Culture: Five Strategies to Keep Women in Tech*, (Washington, DC: Accenture Research: 2019), 3, https://www.accenture.com/content/dam/accenture/final/a-com-migration/pdf/pdf-134/accenture-a4-gwc-report-final1.pdf.

2. Mbula Schoen, "Do Recent Layoffs Mean the Tech Talent Crunch Is Over?" *Newsroom Q&A* (blog), *Gartner*, March 7,

2023, https://www.gartner.com/en/newsroom/press-releases/2023-03-07-do-recent-layoffs-mean-the-tech-talent-crunch-is-over.

3. Accenture Research, *Getting to Equal 2020: The Hidden Value of Culture Makers*, (Washington, DC: Accenture, 2020), 9, https://www.accenture.com/ae-en/about/inclusion-diversity/_acnmedia/Thought-Leadership-Assets/PDF-2/Accenture-Getting-To-Equal-2020-Research-Report.pdf.

4. Accenture Research, *State of Cybersecurity Resilience 2021: How Aligning Security and the Business Creates Cyber Resilience*, (Washington, DC: Accenture Research, 2021), 8, 9, https://www.accenture.com/_acnmedia/PDF-165/Accenture-State-Of-Cybersecurity-2021.pdf.

5. Jonathan Reed, "The cost of Data Breach for Government Agencies," *Data Protection* (blog), *Security Intelligence*, September 7, 2022, https://securityintelligence.com/articles/cost-data-breach-government-agencies/.

6. Anthony Martinez, Sheridan Christnacht, "*Women are Nearly Half of US Workforce but Only 27% of STEM Workers,*" America Counts: Stories (blog), *United States Census Bureau*, January 26, 2021, https://www.census.gov/library/stories/2021/01/women-making-gains-in-stem-occupations-but-still-underrepresented.html.

7. Sapana Cheryan, Allison Master, Andrew Meltzoff, "*There are too Few Women in Computer Science and Engineering,*" Opinion (blog), *Scientific American*, July 27, 2022, https://www.

scientificamerican.com/article/there-are-too-few-women-in-computer-science-and-engineering/.

8. Accenture and Girls Who Code, *Resetting Tech Culture: Five Strategies to Keep Women in Tech*, (Washington, DC: Accenture Research, 2019), 13, https://www.accenture.com/content/dam/accenture/final/a-com-migration/pdf/pdf-134/accenture-a4-gwc-report-final1.pdf.

9. Accenture and Girls Who Code, *Resetting Tech Culture: Five Strategies to Keep Women in Tech*, (Washington, DC: Accenture Research, 2019), 23, https://www.accenture.com/content/dam/accenture/final/a-com-migration/pdf/pdf-134/accenture-a4-gwc-report-final1.pdf.

10. Accenture and Girls Who Code, *Resetting Tech Culture: Five Strategies to Keep Women in Tech*, (Washington, DC: Accenture Research, 2019), 17, https://www.accenture.com/content/dam/accenture/final/a-com-migration/pdf/pdf-134/accenture-a4-gwc-report-final1.pdf.

11. Cecilia Kan, "Google Data-Mines Its Approach to Promoting Women," Tech Policy (blog), *The Washington Post*, April 2, 2014, https://www.washingtonpost.com/news/the-switch/wp/2014/04/02/google-data-mines-its-women-problem/.

12. Accenture Research, *Equality = Innovation: Getting to Equal 2019: Creating a Culture That Drives Innovation*, (Washington, DC: Accenture, 2019) 4, https://www.accenture.com/content/dam/accenture/final/a-com-migration/thought-leadership-assets/accenture-equality-equals-innovation-gender-equality-research-report-iwd-2019.pdf

13. Accenture Research, *Equality = Innovation: Getting to Equal 2019: Creating a Culture That Drives Innovation*, (Washington, DC: Accenture, 2019) 3, https://www.accenture.com/content/dam/accenture/final/a-com-migration/thought-leadership-assets/accenture-equality-equals-innovation-gender-equality-research-report-iwd-2019.pdf

14. Beth Almeida, Isabela Salas-Betsch, "*Fact Sheet: The State of Women in the Labor Market in 2023,*" *Building an Economy for All* (blog), *Center for American Progress*, February 6, 2023, https://www.americanprogress.org/article/fact-sheet-the-state-of-women-in-the-labor-market-in-2023/.

15. Lisa Nagele-Piazza, "*The Importance of Pay Equity,*" HR Magazine (Spring 2020), https://www.shrm.org/hr-today/news/hr-magazine/spring2020/pages/importance-of-pay-equity.aspx.

16. Amy Novotney, "*Women Leaders Make Work Better. Here's the Science Behind How to Promote Them,*" *Women and Girls* (blog), *American Psychological Association*, March 23, 2023, https://www.apa.org/topics/women-girls/female-leaders-make-work-better.

17. Alice Eagly, "*Women as Leaders: Leadership Style Versus Leaders' Values and Attitudes,*" (Boston, MA, Harvard Business School, 2013), https://www.hbs.edu/faculty/Shared%20Documents/conferences/2013-w50-research-symposium/eagly.pdf.

18. Amy Novotney, "Women Leaders Make Work Better. Here's the Science Behind How to Promote Them," Women and Girls (blog), American Psychological Association, March 23,

2023, https://www.apa.org/topics/women-girls/female-leaders-make-work-better.

19. Katelyn Fossett, "A 'Watershed Year' for Women on Corporate Boards," *Politico*, April 15, 2022, https://www.politico.com/newsletters/women-rule/2022/04/15/a-watershed-year-for-women-on-corporate-boards-00025478.

20. Monster editorial team, "*7 Ways You Can Support Women in Leadership*," *Workplace Diversity and Equity* (blog), *Monster*, access date 10/16/2023, https://hiring.monster.com/resources/workforce-management/diversity-in-the-workplace/supporting-women-at-work/.

21. Ibid.

22. Ibid.